S60 SMARTPHONE QUALITY ASSURANCE

S60 SMARTPHONE QUALITY ASSURANCE

A Guide for Mobile Engineers and Developers

Saila Laitinen
Nokia, Finland

John Wiley & Sons, Ltd

Email (for orders and customer service enquiries): cs-books@wiley.co.uk
Visit our Home Page on www.wiley.com

Other Wiley Editorial Offices

John Wiley & Sons Inc., 111 River Street, Hoboken, NJ 07030, USA

Jossey-Bass, 989 Market Street, San Francisco, CA 94103–1741, USA

Wiley-VCH Verlag GmbH, Boschstr. 12, D-69469 Weinheim, Germany

John Wiley & Sons Australia Ltd, 42 McDougall Street, Milton, Queensland 4064, Australia

John Wiley & Sons (Asia) Pte Ltd, 2 Clementi Loop #02-01, Jin Xing Distripark, Singapore 129809

John Wiley & Sons Canada Ltd, 6045 Freemont Blvd, Mississauga, ONT, L5R 4J3, Canada

Wiley also publishes its books in a variety of electronic formats. Some content that appears in print may not be available in electronic books.

Anniversary Logo Design: Richard J. Pacifico

British Library Cataloguing in Publication Data
A catalogue record for this book is available from the British Library

ISBN 978-0-470-05685-1 (PB)

Typeset in 11/13pt Zapf Humanist 601 by SNP Best-set Typesetter Ltd., Hong Kong.
Printed and bound in Great Britain by Antony Rowe Ltd, Chippenham, England.
This book is printed on acid-free paper responsibly manufactured from sustainable forestry in which at least two trees are planted for each one used for paper production.

Table of Contents

About the Author

Saila Laitinen, an engineer, grew up and went to school in Oulu from where she moved to the capital area of Finland in 1995. She is married and has two children. She graduated from Oulu University, from where she gained an M.Sc. in computer engineering.

She joined Nokia in 1995 and has worked in a variety of positions and organizations within the company since then. She started in Nokia Networks and worked as a software engineer for three years. This gave her a solid base including an overall understanding of software and development challenges as well as network technologies. During those three years she also worked as a project manager on two software projects. This provided her with first-hand understanding and know-how of how challenging it is to run a project both to budget and to a given schedule.

After Networks Saila joined the Nokia Ventures Organisation to lead testing activities in one venture. Then she gained international experience and worked in Nokia Hungary as an expatriate. In Hungary her responsibility was to manage the overall testing activities of Nokia's very first presence server-product. After Hungary she moved back to Finland and joined the S60 product line, where she led the testing, triage and technical consultancy teams. These teams worked on a daily basis in response to S60 customer products and provided them with platform expertise in different technologies and activities. Lastly Saila joined Forum Nokia where she currently leads the global consultancy function to serve the biggest developer community innovating on top of Nokia platforms.

Working in all these organizations and both in Finland and abroad has given her very clear insight into mobile technologies and cultural differences. Several years experience working with S60 customer programs has given her the first hand knowledge needed to write this book.

In addition to this book she has publications in different testing and quality assurance related conference proceedings.

Preface

The idea of writing this book came to me in 2002. Since then it slowly matured into a state where I knew exactly what I wanted to write and finally into an agreement with the publisher John Wiley & Sons Ltd. S60 is without any doubt the world's leading smartphone platform and it is indeed a remarkable one. My dream and target is to help customers to write programs that create mobile phones on top of the S60 platform and to help them understand and see the huge value of it to their businesses. This book explains the S60 platform, platform-based device programs and quality assurance factors for such programs in sequence, so that the reader gets an understanding of how to prepare and organize the development of a platform-based mobile device. I have tried to share my experience and the way I felt while working on S60 so that the reader can discover the fascination of smartphones and also be prepared to handle the most demanding issues and risks in his project.

The book consists of four parts. The first part starts by comparing the smartphone concept with the feature phone. The smartphone is explained naturally through S60 and its architecture. S60 architecture consists of a cellular modem controlled by modem software, the Domestic Operating System (DOS), and an application processor engine controlled by Symbian OS and S60 software. All these parts are explained to the reader so as to give a comprehensive understanding of the main S60-based device building blocks. In addition I have explained two of the most important challenges in implementing an S60-based mobile phone, Binary Compatibility and Certificates.

The second part concentrates on quality, what it means, how to gain it and what the pitfalls are in gaining the required quality for different product programs. Quality can mean different things to different people. The meaning also varies between products. However, the one and only common element of quality is the way the consumer or customer sees the product and how well it fits its use.

The third part explains the most common stumbling blocks in implementing a high-quality product, with special attention naturally being given to an S60-based phone program.

The fourth part explains the tools used to tackle the challenges that end with a product with very few errors in the marketplace. This part starts by introducing testing as a tool to show how far a program is in its quality targets. Testing alone never increases quality, it only makes it visible. To understand a product's quality state makes it easier to understand how much work is still needed before the product can be shipped. Increasing quality equals fixing existing defects. Fixing the right defects is one thing but another one is the timing of the fixes. Both of these elements are introduced in the fourth part.

Acknowledgements

Writing this book unaided would not have been possible at all. It is the result of the very best teamwork one could have. Luckily I have had a tremendous team to support me every step of the way. This team not only provided me with lots of material but also generated new material whenever needed. Even though I might have forgotten someone, I would still like to highlight the following as part of the critical chain behind this book: Samuli Paavola, Antti Saukko, Veli Sertti, Sandor Szilagyi and Mirkka Ylisuvanto. Samuli's skills and know-how in error handling enabled me to explain the importance of proper error management as a success factor in a mobile device program. Antti brought his invaluable and professional platform and tools knowledge that helped me with related parts in this book. Veli gave his expert advice on certification related parts. Sandor provided his competence beyond comparison to the binary compatibility related parts of this book. Mirkka helped me understand testing processes as well as other platform-specific testing-related topics so that I was able to explain them correctly.

To really understand the issues customers face in their projects requires first-hand experience. This was given to me by Heidi Melen and Kalevi Ratschunas. I was honoured to work side by side with them in one of the Licensee's device programs. That program opened my eyes to the challenges that customers have in creating a phone with S60 on it.

I also feel the need to give my deepest thanks to my employer Nokia. After being privileged to work for this company for over eleven years, I appreciate everything this company has given me. It has given me so much, usually everything I have ever asked for:

enough challenges to keep me motivated, enough support and rest to encourage me when tired and during difficult days plus enough feedback to become better than I was. I have been fortunate to see this company from many angles, from the purest R&D work to customer support and strategy work. And I am still humbly looking forward to all the years to come that hopefully we will share.

Last but not least I need to mention three main supporters of mine, my husband Hannu, daughter Veera and son Lauri. Hannu has given me so much inspiration by showing me that one can get results only by having a hard working mentality. Since this book is a personal project, I have written it in my spare time. This has unfortunately meant unforgivable neglect of my most loved ones, Veera and Lauri. While writing this during weekends, evenings, nights, early morning hours and holidays, you two asked me whether I would ever have time for you? I am now happy and relieved to tell you, this project is finally completed and I am all yours!

Chapter 1: Introduction to S60

It can be surprising to realize how complex a device a mobile phone really is, and how difficult it is to create one. Because of that, it is not at all surprising to see how difficult it is for any manufacturer to succeed in the mobile phone market. The purpose of this chapter is to describe the tip of the iceberg of why that is so, by describing the elements of a typical smartphone from a logical architecture point of view. Later chapters will go into further detail about creating an S60-based device. The general architecture of an S60-based smartphone consists of a cellular modem controlled by the modem software, the Domestic Operating System (DOS) and the application processor engine controlled by the Symbian Operating System (OS) and S60 software.

What is it that makes a device a smartphone? The simplest mobile phone (Figure 1-1) enables voice calls and short messaging (SMS). In addition, a contact list can be considered as a fundamental feature of any mobile device. The next step from 'any' device is a feature phone, which contains some significant additional functionality:

- calendar for keeping track of appointments
- a web or WAP (Wireless Application Protocol) browser

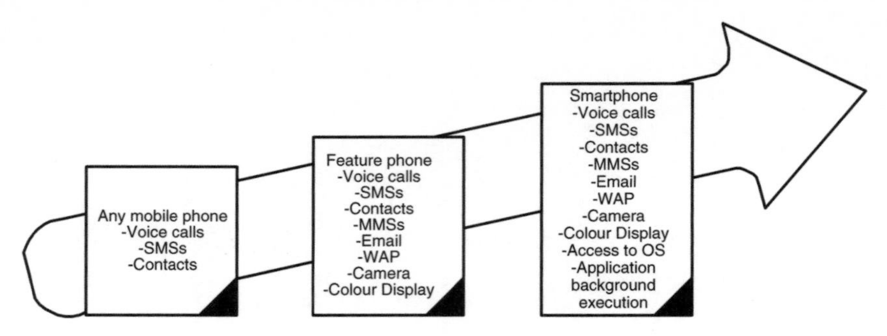

Figure 1-1. Mobile phone evolution.

- multimedia messaging support (MMS)
- email
- games and other pieces of application software
- a colour display
- a digital camera
- support for additional memory through the use of a memory card
- etc.

A feature phone has a relatively simple, but effective, proprietary software environment based on a real-time operating system (RTOS). Smartphones, on the other hand, use a more advanced, open high-level operating system with support for multitasking, expandability, multimedia or convergence features, application interaction and so on.

Feature phone functionality may have support for additional extensibility through installable software applications, usually based on the Java 2 Micro Edition (J2ME) technology and the Java programming language from Sun Microsystems. Smartphones, while supporting J2ME, also support software development through direct, native access to the underlying operating system and its functions (through, for example, software written using the C++ programming language). Perhaps the most notable difference, however, between a smartphone and a feature phone is the way the applications use the phone resources. In feature phones only one application can be run at any given time, whereas in a smartphone the execution of multiple applications happens in the foreground (visible to the phone user on the display) or in the background, and all the applications can access phone or operating system resources simultaneously, including other applications and network services.

1.1 The Competitive Advantage of the S60 Platform

The S60 Platform is the world's leading smartphone software platform, offering a feature-rich software base for phones with advanced data capabilities. It includes the Symbian OS and the Nokia S60 UI (user interface). This UI is the most extensively researched and thoroughly developed graphical user interface (GUI) ever created by Nokia. Its inclusion in the S60 Platform ensures UI consistency across all phones based on the S60 Platform from all device manufacturers. The S60 UI is designed for one-hand operation of advanced and consumer-friendly data services. It supports a variety of different functions, including two softkeys, five-way navigation and an application launching and swapping key, as well as *Call creation* and *Call termination* keys. To improve and facilitate text input, it includes a *Clear* key and an *Edit* key. In addition, it uses the standard 12-key number keypad with alpha printing.

S60 now includes scalable UI support for the following screen resolutions (in pixels):

- 176 × 208 (classic)
- 240 × 320 (QVGA)
- 352 × 416 (double)

Scalable UI also supports each screen resolution in either portrait or landscape view and introduces a scalable graphics (SVG) format for icons and themes.

In addition to the quality assurance of an S60-based phone, this book guides the reader through the concept, idea and competitive advantage of S60 in the global smartphone markets. Nokia's Mobile Software (MSW) is the organization behind the S60 platform. The Product Creation Community (PCC) members represent the leading third-party companies in different regions when it comes to manufacturing a mobile phone. They get the full S60 release at the same time as the device programs, and are entitled to use it for internal competence development purposes. There is a Developer Community of developers around the world who are innovating on top of the Nokia platform. A commonly used description for all these is the S60 ecosystem. The entire S60 ecosystem is shown in Figure 1-2. This licensing model enables the platform to be used in different manufacturers' device programs.

In addition to the platform itself, MSW works on a reference hardware that contains the platform as well as modem software. Developers of customer programs can buy and utilize this pre-integrated product as a base for their final product. The usage of the reference hardware is highly recommended as it provides a half-ready product and allows the developer to dedicate resources to differentiation only.

The term 'Licensee' in this book can mean either a Nokia device program or another manufacturer's device program. Another name that is used in this book for the Licensee is customer program. All customer programs are treated equally. In practice, this means that all of them have equal access to all platform releases and documentation.

The Product Creation Community (PCC) is composed of technology integrators and other companies competent to participate in the customer product program of making a phone. PCC companies can provide help to Licensees in platform integration, testing and development activities, just to mention a few. S60 Product Creation Community members are divided into four categories:

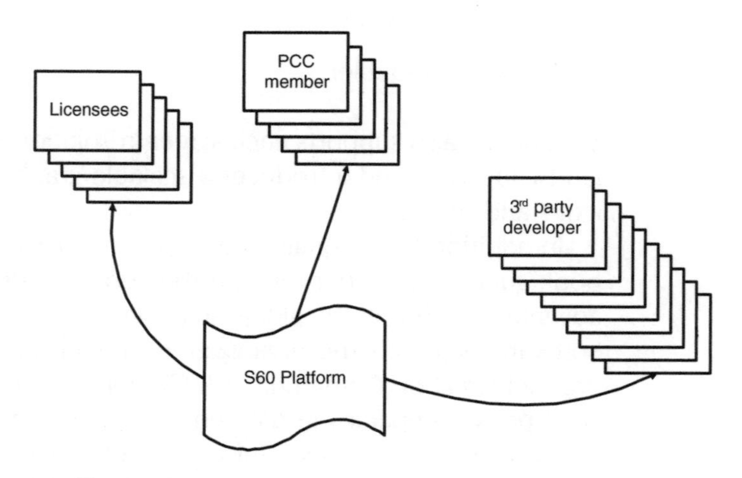

Figure 1-2. The S60 Ecosystem.

- **Boutiques** – experts in designing complete S60 phones and managing entire S60 phone projects
- **Competence Centers** – top-tier software companies with deep S60 end-to-end understanding and extensive S60 project support
- **Wireless Technology Providers** – experts on the hardware platforms or hardware components upon which S60 phones are built
- **Contractors** – skilled software companies offering focused expertise in specific technology areas

Each member is carefully selected and required to meet stringent qualification criteria.

The third-party developer community represents the biggest entity in terms of the number of participants. Forum Nokia is the entity in Nokia that supports these 2.5 million developers worldwide. It arranges training throughout the world, manages the discussion board on technical topics and provides case-based technical support for independent issues as well as tailored technical consultancy for customer projects. Developers can implement applications on top of the S60 platform by utilizing both JAVA and C++ interfaces. Developers make their profit by selling these applications. Together these applications represent one of the widest mobile application portfolios in the world.

The ecosystem is like a chain with equally important pieces. Together they create a unique and strong base for a special competitive advantage amongst platform providers. If one piece breaks, the entire chain is paralysed.

1.2 S60 Architecture

S60 consists of numerous architectural units, for example the Symbian OS, the Domestic OS adaptation and UICON. This section explains in turn the platform's main building blocks and their purpose. Other important concepts are also briefly introduced below. The overall architecture of a smartphone is introduced in Figure 1-3.

Once a customer program receives a platform release, it needs to integrate it into the hardware and the Domestic OS. Base Port is the exercise of adapting the Symbian kernel to a particular hardware. Kernel port consists of providing the Symbian kernel access to the necessary hardware functionality. Symbian runs in the following two modes:

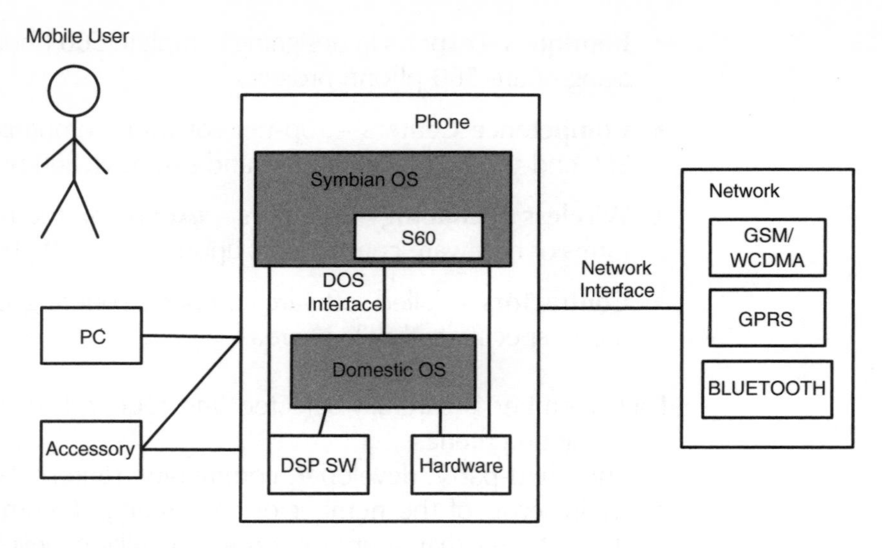

Figure 1-3. Smartphone architecture.

- **User mode** – kernel services can only be accessed through the EUSER.DLL. The lack of a proper kernel port does not stop the development of user-side components because the platform provides a complete Kernel Port for the PC environment under the Windows Operating System. This is called the emulator.
- **Kernel mode** – EUSER.DLL is an interface between common code and hardware-specific code. In the other words, kernel mode means that the software is run on the target hardware.

1.2.1 The Symbian Operating System (Symbian OS)

S60 is based on the Symbian operating system, which provides several services to the platform and to platform-based devices. Such services are, for example, the User Interface (UI), applications and middleware.

1.2.2 Domestic Operating System (DOS)

The Domestic Operating System (DOS) is the proprietary operating system and no interfaces in it are open to third-party developers. DOS plug-ins are device specific and need to be implemented by the customer program.

1.2.3 User Interface (UICon)

S60 includes the user interface components needed by an application. UICon is a graphical user interface (UI) library for reference-design (DFRD) independent functions based on EIKON, which is the original graphical user interface library for the Symbian OS. Use of such components guarantees the implementation of the application of the user interface by developers.

1.3 Summary

This chapter has briefly introduced the smartphone, what it is, its architecture and how it differs from other device types. The basic components of the S60 Symbian operating system, the domestic operating system adaptation and the user interface library are all explained in this chapter. The overall architecture of the S60 consists of the Symbian Operating System, the User Interface components and adaptation to a device-specific Domestic Operating System plus telephony software. The S60 ecosystem consists of Nokia's Mobile Software, platform Licensees (device manufacturers), the product creation community and third-party developers, which together provide a strong basis for the success of the platform.

Chapter 2: Selecting the Baseline

The software industry has had to adapt to a very new mindset since the early 1990s. Software began to play a key role in very many products after consumers had started to appreciate the ever-growing number of new features in these products. As a result of the new features and technologies, the average size of the software in a mobile phone has grown quite a lot. This increase is not only due to the new complex functionality (often described as 'digital convergence') required in these new products, but also because of the need to put more structure and discipline into the software system in order to make it more controllable. Well-known features such as modularity, scalability and decoupling form part of this. Engineers are also facing challenges in introducing an operating system on the signal-processor side, in order to be able to meet new demands.

S60 Smartphone Quality Assurance Saila Laitinen

2.1 Manny Lehman's Law

As a program evolves, its structure will become more complex. Just as in physics, this effect can, through great cost, be negated in the short term.

Michael W. Godfrey and Qiang Tu based their case study on the evolution of open-source software on Manny Lehman's law:

When a software system gets bigger, its resulting complexity tends to limit its ability to grow. As an advice to this; the complexity needs to be well managed and maybe even the entire system needs to be redesigned every now and then.

MSW releases the S60 platform at a very early stage in the development of a platform program. The maturity and stability can therefore be quite unreliable. These early versions are described as being of research and development (R&D) quality. The MSW follows an incremental process in developing the S60 platform. In practice, this means that the program increases the maturity of one feature at a time before anything else is included into the release.

In practice, the development process of a platform, like that of any similar sized software product, is a bit more complicated and not as straightforward as this sounds. The platform program procedure includes elements from both iterative and incremental development processes. This can be seen in two ways:

- the overall increasing maturity throughout the program, which indicates that iterative process are being followed
- the sequential development of the main features, which clearly indicates that an incremental process is being used

As shown in Figure 2-1, each feature is considered as an individual sub-system. This sub-system can only be verified when other surrounding sub-systems are available. In order to minimize the number of stubs and drivers needed during testing, both implementation and verification orders need special planning and scheduling at an early enough stage.

As an example, release 2.8 introduced the Scalable UI as a new feature. It has been structured so that the layout information cannot just be hard-coded anywhere. Instead, the Conversion Description Language (CDL) interfaces, which allow access to the layout data

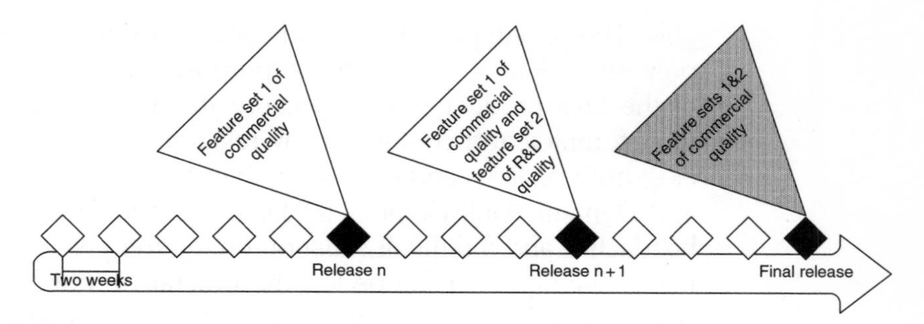

Figure 2-1. Release cycle.

based on the Look-And-Feel (LAF) specifications, enables it to be stored. Since this applies to all applications, the implementation order of the sub-systems should be the following:

1. Scalable Vector Graphics Tiny
2. Core Services
3. UI Framework Core
4. UI Framework Extensions

After all the sub-systems are in place, the implementation of other sub-systems or applications using the scalable UI feature can begin.

2.2 What is so Challenging about Selecting the Best Baseline?

The baseline in this context means the bi-weekly release that is the most recent to be fully integrated in the customer program as a complete platform. As simple as this sounds, figuring out which out of many releases should be treated as the baseline for the entire system is not a very simple task. The program needs to balance two things, time and maturity. The earlier the baseline that is selected, the higher is the probability of shipping the product on time. However, there are tradeoffs. The earlier the baseline, the more unstable it is. The customer programs then receive even earlier versions of S60, in which the maturity of the scalable UI-related sub-systems is debatable. The phone program should consider using these early versions only for evaluation and rehearsal purposes and not for manufacture of the final product. This is because the architecture of the S60 is very complex and contains many relations and internal dependencies, both obvious and obscure, within the sub-system components and Dynamic Link Libraries (DLL). These relations can cause many headaches and a difficult situation during the period while they are fixed. During this fixing period the program is obliged to build the system from bits and pieces, i.e. 'gluing' together some components and files from week *x* release, some components from week *x* + *y* release etc.

On the other hand, if the program waits until the platform has reached the level of commercial quality and there are no implementations by the manufacturer available in the mean while, it may miss the market window. Deciding this without losing out either way requires both deep technical and business knowhow.

2.3 How should the Baseline be Selected?

Once the baseline has been selected, in later versions newer parts only replace separately selected parts of the release. This selection is one of the key decisions in the S60-based phone program. The decision has a significant impact on the success of the whole project. It is good to include the following in the preliminary decision work:

1. The maturity of the platform – the maturity should be high enough to avoid an unnecessarily large amount replacement of components and sub-systems in the program.
2. Maturity of the Licensee's own implementations – the maturity should be high enough to avoid an unnecessarily large amount of correction in the interfaces between the platform and developers' own components.
3. Least stable sub-system – the least stable sub-system/feature can be the one with the lowest risk of affecting other parts of the system when it is fixed. If the least stable sub-system or feature has several dependencies on other parts, every fix on it may damage functionalities that are already working. If, on the other hand, it is a relatively 'independent' component/sub-system, the low maturity (= quite a lot of fixes yet to come) should not be too risky.

4. Program timing – the program should choose the baseline early enough to avoid missing the markets window by providing an outdated product to the market. This can force the product program to include additional features by porting from later versions of the platform, which of course is not desirable.

2.3.1 Baseline Maturity

Every platform release goes through Basic Acceptance Testing (BAT) before it is shipped. Each release also contains the results of the BAT round. These results indicate the overall functional stability of that release. It is highly recommended that the BAT pass rate of the chosen baseline release is high enough to minimize the work of integrating the individual fixes.

2.3.2 Customization Maturity

The customer program naturally wants to include some customization so as to make the final version look more like their own product. This can be done in many ways, some fast to implement and others difficult, some without risk and others very risky. Below are introduced three of many possible examples of customization:

- Adding one's own or third party applications/features on top of the platform represents relatively simply customization. It is the safest way to customize as long as the changes in the platform are very limited and well controlled.
- Another option, with a greater degree of freedom, is also to customize the UI on the platform side. This means that the platform needs some modification. If the program manages to make such modifications wisely, the risk is manageable.
- The most demanding customization activities are those in which some existing features/technologies are to be removed from the platform. This should be done very carefully and in a controlled way in order to not to affect any remaining functionalities.

An example: The Licensee program decides to remove Bluetooth (BT) from its product. This can be done in two ways, either by removing the BT enabler implementation or by muting the BT. Although the Bluetooth implementation in the platform shares many resources with other connectivity options such as infrared and USB, removing BT needs to be done in a prudent way. If it is done unprofessionally, other connectivity types can be disturbed.

2.3.3 Least Stable Sub-system

When choosing the baseline, the program should be fully aware of the least stable sub-system and its importance for the entire product. It does not matter whether it is on the platform side or in the licensee's own implementations. What matters is whether it implements a critical interface, i.e. between the platform and the licensee's own customizations, or otherwise has several dependencies on other parts of the software. As this is the weakest link, it is very probable that most fixes will take place in that part. If one expects to have relatively many changes made to the intermediate components of the software, it is good to be aware of them and prepare the project organization to tackle the need for major changes.

2.3.4 Program Timing

That the early bird catches the worm applies to the mobile phone industry. Therefore, choice of the baseline has a natural link to the success of the program. In the other words, nobody wants intentionally to sell out-dated terminals with a feature set that has been introduced on other available phones several months earlier. In addition, operators' requirements are very tough and they would like to see most, if not all, features included in all terminals sold via their sales channels. If the program waits for the platform version to be of commercial quality level (i.e. its quality has reached the level that was business-wise clever enough to stop further maturing) and chooses that to be the baseline, consumers may pick a competitor's product with a more complete feature set. The earlier the program chooses the baseline the better, as long as maturity-related aspects are also considered and analysed.

2.4 Summary

In all complex multi-supplier software programs, the software baseline selection is the core of the matter. It is certainly one cornerstone of the program and therefore it is vital that the architects control it. This chapter provides detailed information on the special challenges that each S60 customer program faces when making this important decision. The S60 release cycles are described, as well as the potentially contradictory facts when the critical choice of the baseline is made. This chapter tries to provide an overview of proper baseline selection, by introducing the most common pitfalls in the process. Selecting the best possible baseline is without doubt one of the cornerstones in having a successful S60 product, while it is one of the trickiest decisions for the program to make.

Chapter 3: Release Management

A phone program normally involves several software suppliers, as shown in Figure 3-1. Release management can be very demanding in such a multi-supplier environment. Internal teams can also be considered suppliers if such teams do not communicate and interact with each other in a daily basis. Scheduling and mapping the inflow of different sub-system releases and combining them together, can turn out to be one of the biggest risk factors in a program. Therefore release management requires very strict processes and policies as well as everybody's commitment to follow them. A non-analyzed risk in one sub-system maturity can have tremendous impact on the program success. The challenging variables having direct impact on the release management are for example overall complexity of an architecture and software, size of a software system (number of lines in code), estimated number of individual fixes accepted to be integrated after code complete and number of used suppliers.

Let us start by taking the above examples one by one. Software complexity can be divided into two aspects, architectural complexity

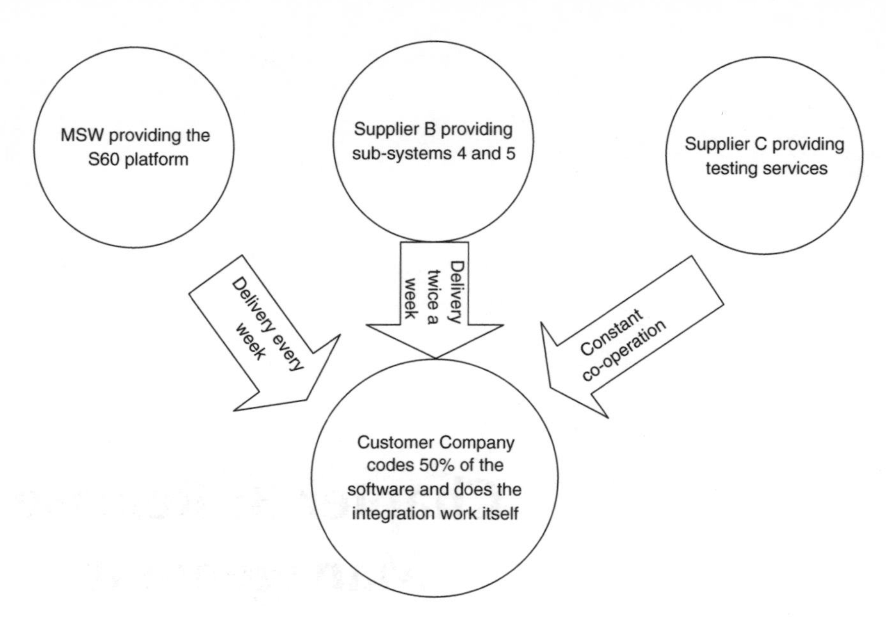

Figure 3-1. Multi-supplier program architecture.

and code complexity. Architectural complexity has become an issue, especially concerning Object Oriented (OO) software where the importance and value of software reusability, maintainability and adaptability to small chips (as in the mobile phone industry) has been recognized. Unfortunately, the drawback of this is that managing shared resource files can be a huge task, especially in a defect-fixing mode. Code complexity can be measured, for example by using McCabe's cyclomatic complexity formula, which measures the complexity of a particular function by checking the number of branches in the code. A method with no branches has a complexity of one; a method with one branch has a complexity of two, etc. There are tools available that can be used in measuring the complexity of particular functions/code.

Both of these complexity aspects have an impact on the importance of release management in a program.

System size is in most cases measured in terms of the Lines of Code (LOC) measurement. LOC is probably the clearest metric for indicating size, but it is not necessarily the most useful one. System size can also be exemplified by the number of sub-systems, components or Independent Software Vendors (ISV) used in the program. These measures can often be more controversial than the simple LOC, but effective use of them in release management can be worth every penny.

Nobody can turn dross into gold. The estimated number of acceptable fixes that can be integrated into the baseline is a good metric to keep in mind. The more fixing that is needed, the worse the original and resulting codes. It has also been said that one fix creates five new defects in the system. Some hints on deciding what to fix and what not is discussed in more detail in later chapters.

In order to guarantee equal access to the software plus the possibility of following the overall maturity development of the platform, MSW releases a new platform package every two weeks to all customer programs. This means two deliveries every month for each incremental release. If it wishes to do so, a customer program can integrate each new software package every two weeks into its own development environment, but that may cause a significant amount of work as well as increasing the complexity of reverse engineering if regression is necessary. Whether integrating everything in each release is worth doing depends on the number of modifications made to the platform as well as to other procedures in the program, but it certainly adds extra managerial challenges to the building process. The following chapters provide different viewpoints on how to evaluate which releases should be integrated as a complete package and which may be worth neglecting completely.

3.1 The Build Cycle

First, a customer product program needs to define which platform increment fulfils most of the program's expectations. This definition requires a knowledge of resource availability and a wide understanding of the markets, as well as deep expertise about the technologies. After all, this is always going to be a business decision. The evaluation phase can take months and meanwhile the markets can change.

It is easier to follow the progress of a program if enough indicators have been identified for describing its status. One quite widely accepted way is to use specific milestones with corresponding identifiers, as shown in Figure 3-2. Each such milestone is introduced below:

- Project initiation (L-1):
    - S60: program project manager nominated
    - Customer: business case (high-level)

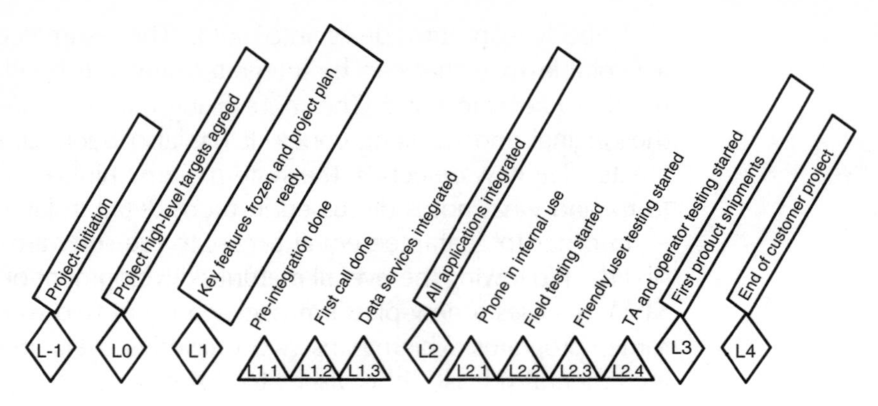

Figure 3-2. An example of phone program milestones.

- Customer and S60: key features and target market segment identified (overall schedule – time-to-market)
- Customer: main feature set decision made
- Customer and S60: key suppliers defined (dependencies)

- Project high-level targets agreed (L0):
  - Customer: product program business-case feasibility study approved
  - Customer and S60: critical chain understood and back-up plans made so that realistic targets can be set to balance the business requirements and the capability
  - Customer and S60: customer architectural specification analysis carried out and reviewed
  - Customer and S60: first program plan established (from L1 to L3 including resource commitments in place)
  - Customer and S60: suppliers and PCC members identified
  - Customer and S60: S60 licensing agreement ready
- S60: Project team nominated
- Key features frozen and project plan ready (L1):
  - S60: release and support plan established
  - S60: readiness to deliver pre-releases established

    - S60: product specification frozen and all processes (error management, change management, testing, etc.) agreed with customer
    - Customer and S60: pre-integration done (L1.1)
    - Customer and S60: Stand-alone applications integrated on the top of the Baseport (target device running in PDA mode), S60 starts from shell, communication between systems works at basic level
- First call done (L1.2):
    - Voice communications work (or first connection to the network; can be also sms, data, etc.); additional features such as camera working
- (L1.3):
    - CSY pre-integration done
    - second-level connections work (browsing, mms, mail), data services integrated, all the remaining applications integrated (L2)
    - integration completed
    - code completed
    - customer code line management planned
    - error management taken into use
    - all S60 and third-party applications and features integrated and verified; localisations ready
    - full BAT rounds can be started
    - full system test rounds can be started
    - final L3 targets agreed
- Phone in internal use (L2.1):
    - Proto phone is used as primary phone within project team
- Field tests started (L2.2):
    - testing on different operators' networks started

- friendly user tests started (L2.3):
    - testing with selected internal and partly external end users organized and started
- Type approval and operator tests started (L2.4):
    - type approval testing (e.g. GSM, Bluetooth and Java certifications and qualifications) started
    - testing with selected key operators started
- The first customer product is shipping to market (L3):
    - software maturity in all areas at a commercial quality level
    - all approvals passed
    - preparing for maintenance mode started
- End of customer project (L4):
    - handover to maintenance mode completed
    - Update practices agreed (new versions etc.)

Reaching a milestone can also be used to show appreciation to the employees and thus further motivate them to face the remaining tasks and challenges. As seen in Figure 3-2, the word integration pops up several times. Integration often equals building and therefore the build cycle should already be considered in the evaluation phase of the program before any code has been implemented.

As the platform is released every two weeks, it can be assumed that customers most probably make a new build minimum every two weeks as well (at least at the beginning of the program). In addition, customers most probably do some in-house development of, for example, telephony and adaptation layer software and therefore they may to carry out additional builds between the S60 bi-weekly releases.

Figure 3-3 shows how the build cycle varies during the phone program; at the beginning it can be quite long, in the middle of the program it tends to shorten and at the end of the program the cycle is again very short. What is noteworthy is that, as the program gets closer to shipping, only those parts of the system that need to be re-built are re-built. The organization needs to be aware of the need to check and change the build cycle at any point in the program. In

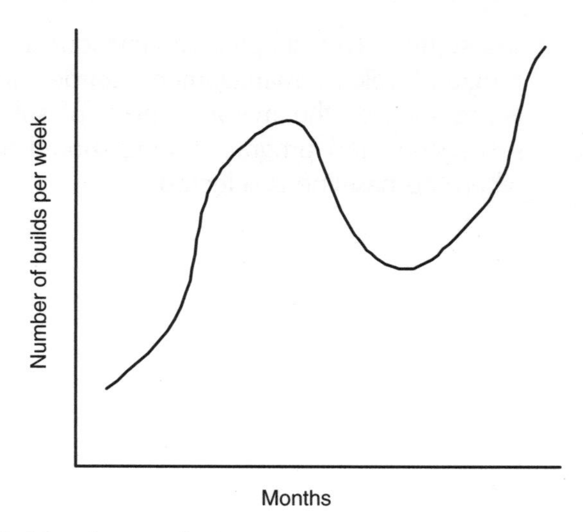

Figure 3-3. Build cycle example.

addition, the overall environment needs to be planned so that it can easily accommodate these changes. Chapter 12 on the build environment goes into more detail.

3.2 Required Testing Activities

As every R&D project is aimed at producing something for the market, there is no time to waste. However, no matter what release is chosen to be the baseline and how long the build cycle is, proper testing needs to be planned and performed. Determining the extent and focus of the testing is an ever-changing process. The very first testing activity with the whole build is called the build sanity check. There is more discussion of the principles of the sanity principals and other issues related to testing procedure in Chapter 8 on platform testing versus platform-based phone testing and the build environment.

3.3 Summary

The more suppliers there are in a program, the greater the complexity of release management. In a multi-supplier environment the program can be affected by the different maturity levels of each sub-system. This chapter has introduced some ways in which the program's release management can be carried out more effectively. Issues such as the maturity levels of sub-systems, the least-stable sub-system and program timing should all be taken into account when the baseline is selected.

Chapter 4: Binary Compatibility

All platform products used for the purpose of further development cannot be described as part of a platform unless they can keep the promise of compatibility. In S60 this means that the platform itself needs to fulfil the compatibility requirements as well as the devices and the applications. In the other words, each device must maintain the platform compatibility and each application developer should respect the public API set up for implementation purposes. The APIs introduced in the present version are expected to be available in the same place with the same attributes and service in later versions. This is called platform compatibility. MSW has made a promise to the entire ecosystem to keep the public API set untouched from release 3.0 onwards. Naturally, the same promise is expected from each product program, and thus each product program must pass the binary compatibility verification before the device is shipped. Although this chapter presents the challenges and the ways in which they can be tackled from a S60 platform perspective, many of the things included here can be copied with slight adjustments to any platform product.

In S60, the architecture itself has been designed so that the compatibility challenges are minimized. What does this mean and how

S60 Smartphone Quality Assurance Saila Laitinen

is it accomplished in practice? First of all, the platform, including all subsystems, is designed to accommodate a layered architecture and, second, the integration of the platform is done in relatively small steps.

The MSW wants the S60 to be widely used and accepted among different terminal manufacturers, end users and developer communities. This is to be achieved by ensuring that, no matter what phone the end user is using, he or she can download applications developed with the platform and use them without any problem in a meaningful way. To minimize any inconvenience in the target, the platform has set tight compatibility criteria, not only for itself but also for the customer programs. The reasoning for binary compatibility can be summarized as follows:

- Operators want to decrease their operational expenditure through portability for their applications, services and content.
- Licensees want to leverage their own and third-party applications in various products.
- According to the business case for third parties, contractors and partners, every new version of an application means additional costs.
- End users want to have applications running, because they have paid for them.
- Licensees want R&D productivity on the basis of stable development interfaces.
- A decrease in the amount of work required for application maintenance on the third-party developer side is desirable.
- Consumers are happier because applications are more portable among devices (switching).

The aim is that, whenever a new S60 platform is released, its components or applications already developed by the phone programs and third-party developers on top of the previous S60 platform version can be reused in their binary form without any alteration. In other words, binary compatibility means that all S60 public interfaces must be supported in upcoming versions of the S60 platform in such a way that there is no need to rebuild executables that are already working and that have been developed and built for some previous version of the S60 platform.

Sometimes, mainly for very important business reasons, the organization can deliberately create binary breaks. This has been seen in 3rd edition of the S60 platform, where a whole new architecture took over from the old one. The third edition of S60 is based on Symbian 9.0 with the implementation of platform security. From version 3.0 onwards MSW is again committed to maintaining the compatibility.

The challenge comes from the fact that each version of S60 (e.g. the second edition feature pack) has its own public SDK, which is made available to the third-party developer community. With every new version there are most probably new APIs included in the public SDK. This naturally contradicts full application compatibility if the application is using some API that was absent from the earlier version and the user tries to use it with a phone that is based on that earlier version.

Again, the nature of the software and the release policy put some constraint on the full binary compatibility requirement. The nature of software is rather special since there is always room for new fixes and some of these fixes most probably also need to take place in the components used in application development. The release policy on the other hand guarantees that the Licensees may use the R&D quality platform for application development. Because of these two factors, there is a need to further specify what binary compatibility really means and how much it covers in the world of S60.

There are two compatibility aspects to the S60 platform: source compatibility and binary compatibility.

Source compatibility means that the application or a client code can be rebuilt without a need to modify the code. Source incompatibilities are introduced for example when;

- The most deprecated APIs are removed – because backward binary compatibility with the S60 2nd Edition cannot be maintained, most deprecated APIs have also been removed from the 3rd Edition, while a number of new replacement APIs have also been introduced. Platform Security and a new application architecture have been introduced.

- The biggest change in the Symbian OS v9.1 and in the S60 3rd Edition is the Platform Security concept. Its main building blocks are Capabilities (set of privileges for applications), Data Caging (secure storage of data), Secure Interprocess Communication (IPC) and memory management. Platform Security also requires a number of changes to the application architecture.

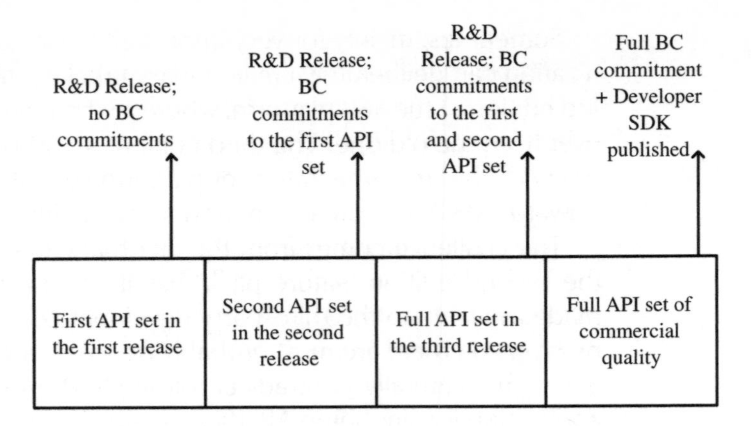

Figure 4-1. API BC commitments.

- A real-time kernel (EKA2) is introduced – EKA2 is the only kernel version supported by Symbian from Symbian OS v9.1 onwards. The compatibility impacts of EKA2 are mainly focused on the need to rewrite device drivers, but there are otherwise a very limited number of source-code breaks.

Binary compatibility means that no rebuild is needed and the application or a client code thus runs on the S60 phone. Potential issues come along when there is a new compiler and tool chain; The S60 3rd Edition introduces new compilation tools (RVCT, GCC EABI), which cause a *full binary break*.

4.1 API Categorization

Both the nature of software and the platform release policy have a significant impact on how easily binary compatibility can be achieved. The platform can never be fully and 100 per cent binary compatible. Therefore, there is a need to have different categories for the APIs to indicate different compatibility commitments. API categorization also helps R&D personnel to understand better how they should treat different APIs.

The APIs still at the R&D maturity stage are not guaranteed to remain the same throughout the program as shown in Figure 4.1. It is already specified in platform specification and design phases which APIs are to be open in the public SDK. The rule of thumb is that, once these APIs are of commercial quality, they can be published to the third-party developer community for application development purposes. This, of course, is not always the complete truth since, if the developers need to wait until the SDP APIs are of commercial quality, they may miss the market window. Therefore, the platform provider should consider publishing an immature (alpha version) SDK to at least a selected number of developers for study purposes. In this way the developer can create the needed competencies before the application implementation starts. When R&D-quality APIs are published to developers, it is made clear that some changes to them are expected and developers should be prepared for these changes.

Sometimes the public API set does not please the developer and the developer needs access to private APIs for its project. Such requests are studied carefully and decisions are made case by case. If the request is approved, the developer needs to understand the risk of using an API that can potentially undergo changes in future releases.

Figure 4-2 shows four main categories of APIs and other interfaces of the platform and platform-based products. Each resource element should find its way into one of the four categories. The APIs in the left-hand boxes have direct impact on the amount of work to be done by each Licensee and Developer, because, if some of the APIs

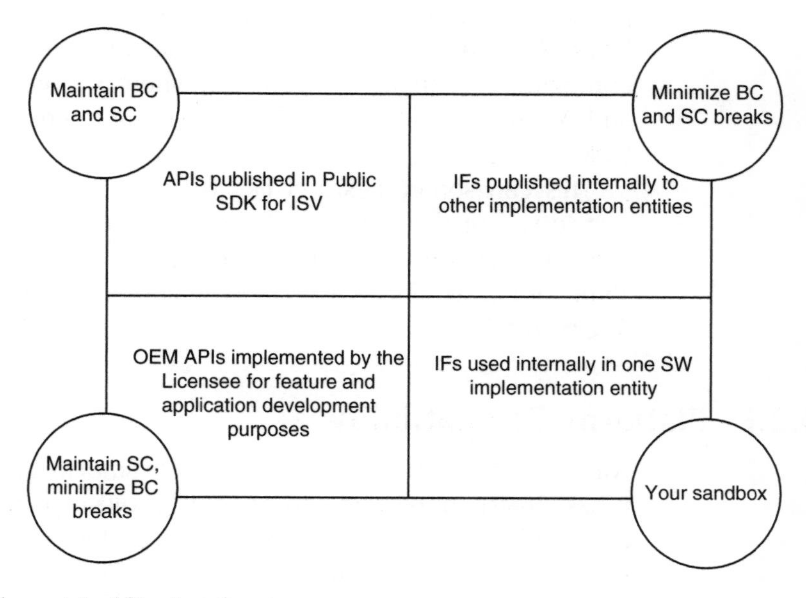

Figure 4-2. API categories.

in these categories are changed, the Developer or Licensee needs to check if he or she needs to make the changes adopted to his or her code as well. To manage the left-hand side requires well-documented design specifications as well as good verification of the APIs to be published. The right-hand side, in contrast, has an impact on the overall stability of the platform. To make the right-hand side stable and to keep things under control, open and effective communication is needed among different the various software entities. This requirement applies to all phone programs equally, while the first requirement is particularly crucial in platform development.

4.2 Maintaining Compatibility

To guarantee the benefits of a unified smart phone base that can be widely used all over the world, several parties need to conform to the existing compatibility criteria. These parties are naturally the platform itself, but platform-based phones and individual application developers also need to commit to follow certain rules. If any of these parties fail in their development, the whole structure could collapse.

Once a new platform version is in the pipeline, the new APIs need to be introduced as extensions to, and not embedded in, the previous releases by changing, for example, the function ordinals in DLLs. All new implementations should be made by following the compatibility scenario in Figure 4-3, where X represents the platform and Y an application developed with the help of the platform SDK.

This can sometimes feel like trying to come up with a jigsaw puzzle where the shape of the pieces is constantly changing. The following sections introduce each player's responsibilities and give some practical hints on how to maintain control in building this 'big picture'.

4.2.1 Platform Compatibility

Naturally, the platform needs to provide stability along with the APIs it has already published and to try to avoid making further changes to them in later versions. Nevertheless, at the same time, some fixes may make it necessary to change these APIs. All such changes need to be clearly communicated to all parties for further analysis if they

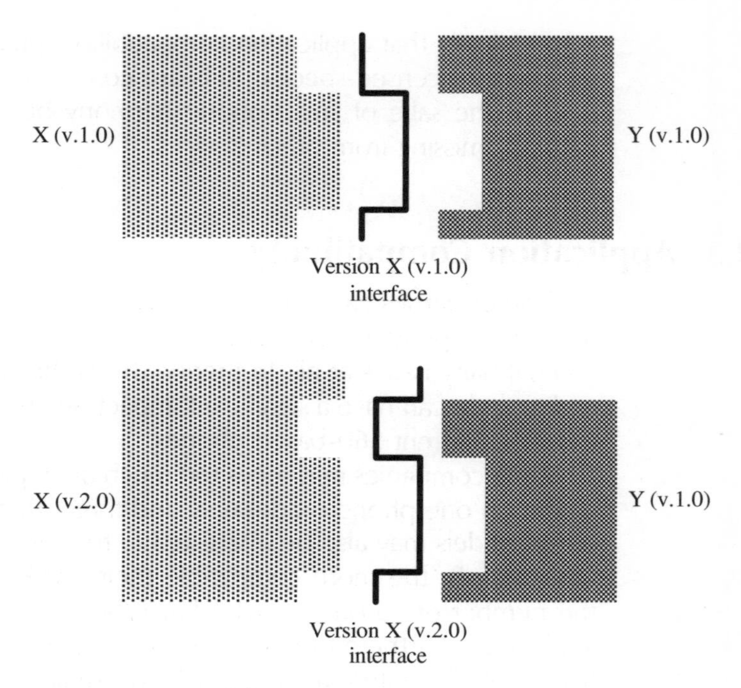

Figure 4-3. Binary compatibility scenario.

need to make any compensating changes to their implementations. Binary compatibility is verified in every platform release and possible breaks are communicated accordingly.

4.2.2 Platform-based Phone Compatibility

No S60-based phone program is permitted to carry out any customisation on these public APIs. All phone programs are informed about such APIs, which have been made open to third-party developers and the immunity of these APIs should be respected. The ability to install and execute third-party applications is ensured by this.

If a licensee wants to publish their own SDK, they should not make available any S60 APIs that have not already been published by MSW. This is because the platform only provides integrity to the public SDK APIs and, if a licensee publishes other APIs for the platform, it will be possible to change them in other S60-based phones as well. Furthermore, the licensee should make possible extensions available very carefully since developers will most probably assume automatically that application compatibility applies on their SDK as well. How Licensee-specific APIs are communicated is very important for the sake of smart-phone harmony because such APIs are naturally missing from other devices.

4.2.3 Application Compatibility

If all other parties carry out their responsibilities, it can be assumed that an independent application developer has no need to consider compatibility issues at all. In theory, this is the case, but in practice coding style can have a significant impact on application compatibility with different S60-based phones.

Some companies may focus too much on implementing the application on one phone model only and forget that consumers having other models may also be interested in the same application. In the other words, the more compatible their application is, the greater the number of consumers who have the possibility of buying their application. Furthermore, some terminal manufacturers publish relatively detailed information on an individual phone model and its capabilities. Together these two things can cause incompatibility within certain applications and this can lead developers to code applications with unnecessary dependencies on a certain phone. An example of such dependencies is, for instance, not making the applications check the state of the phone before carrying out a certain action, the application assuming that all phones have similar states and state transitions. This supports fragmentation and is in contradiction to the overall smart phone platform ideology.

4.2.4 Compatibility Dimensions

As already mentioned in previous sections, in an ideal world all applications work with all phones. This should be the case regardless of what public SDK version was used in application development as well as what platform version the phone uses. Of course, if the application is using some newly introduced APIs, it will not work on phones based on older versions of the S60 platform.

Backward compatibility means compatibility when applications developed in an earlier version of the SDK also work in the later version of the SDK. In this case the later version of the SDK is backward compatible. An application is backward compatible if it is compatible with phones based on older platform versions than that on which the application was developed. In the other words, the application uses no extensions introduced in later versions of the SDK.

Forward compatibility means compatibility if no changes have been made to the SDK APIs of the later versions of the SDK. In this case, the earlier version of the SDK is forward compatible. An application is forward compatible if it runs on terminals based on later versions of the SDK; this is automatically the case if the platform is backward compatible.

4.3 Binary Compatibility Scenario

Let us take a deeper look at to what causes binary breaks and why they occur. As an example, component X (a DLL or a collection of DLLs) presents version 1.0 of an interface to its clients. Client Y version 1.0 is built using X version 1.0. In the next release component X presents a new interface, version 2.0. The question is whether the X version 2.0 is compatible with X version 1.0, i.e. does Y version 1.0 run using interface X version 2.0 or not?

From Y's point of view, the X's interface consists of:

- Header file(s):
  - used by Y at compile time
  - contains X's class and function declarations as well as inline function definitions
  - defines X's set of exported functions (those with IMPORT_C)
- Import Library (X.lib):
  - used by Y at link time
  - contains a list of all functions exported by X
- Exports table (contained in X.dll):
  - used by Y at run time
  - contains addresses of all functions exported by X
- The behaviour of X's exported functions:
  - used by Y at run time
  - implemented in X.dll

1. **CPU Architecture scenario:** The ARM Instruction Set Architecture has six versions. A break is created if the ARM v4 code calls the Thumb subroutine or ARM v4 code returns control to the Thumb subroutine. To prevent this happening, the developer should force the compiler to use ARM Internetworking (ARMI).
2. **Dynamic Linking Library:** Functional ordinal number linking is allocated by either tool chain action or definition files. A break is created if function ordinal numbers change in an uncontrolled manner. To prevent BC breaks, the definitions files should be frozen; if that is not feasible, the additions needed should be implemented by using the 'Ordinal growth and the Extension DLLs' design pattern.
3. **Class Data Members:** Data members represent some property of the class and access to data members is resolved into an offset from the beginning of the object. Binary compatibility is broken if the client contains code that accesses data members directly and the order of the class members has changed since the last version. To prevent breaks, the developer should consider each modification of the data member type, use setters/getters to hide the class structure from clients and not use inline functions becasue they are expanded into client code when the client is compiled.
4. **Class size:** The size of object is determined from the header file at compile time. Binary compatibility is broken if the client contains code that allocates memory for objects and the size of the class has changed since earlier versions. To prevent breaks, the developer should consider each modification of the data member type and use class derivation or the design pattern 'factory method'.
5. **Enumeration:** An enumeration is a distinct integral type that defines named constants. The compiler replaces each enumeration name constant with the corresponding integer number. Binary compatibility is broken if the integer value associated with a named constant is changed in the enumeration. To prevent breaks, the developer should not remove enumeration constants or insert new ones in the middle of the enumeration sequence. A partial solution to overcome the breaks caused by the insertion case is to reserve sequences of values for future use.
6. **Virtual function:** Polymorphism is the ability to process objects differently depending on their class. Binary compatibility is broken if the client instantiates/derives changes in virtual classes and virtual function table compared to the previous versions (for example, virtual function removed, order of virtual functions changed, a new virtual function added). There is no good solution for enabling these kinds of change without breaking the binary compatibility. However, using the design-pattern 'factory method' and making virtual classes non-derivable may help in some cases.
7. **Function signature:** The signature assists the compiler to organize function calls. Binary compatibility is broken if the signature is changed. To prevent breaks, the developer should not modify function signatures. The overload approach should be used instead.
8. **Function behaviour:** Clients may rely on certain function behaviour including a certain set of function input/output values. Binary compatibility can be broken if the developer changes the function behaviour, narrows the set of input values for a function or widens the set of output values for a function.

Binary compatibility can be broken if there is a mismatch between:

- the part of the interface Y built using (header files and import library) and
- the part of the interface Y running using (DLL, containing export table)

In order to provide a deeper understanding of different binary compatibility breaks, eight different scenarios are introduced briefly on page 34. Each of these cases are based on the *Binary Compatibility Theory Training Material*.[1]

4.4 Binary Compatibility Verification

Each S60-based phone must pass a predefined verification set prior to shipping. It is enough to run the tests only once, right before shipping is expected to take place, but if that is the case and the verification reveals a significant number of breaks, the whole phone program schedule will be delayed. It is highly recommended that immediate attention should be paid to possible binary breaks and they should be fixed at an earlier stage. To achieve this, the program should use the verification suite as part of the development work.

4.4.1 The Binary Compatibility Verification Process

MSW provides the binary compatibility test suite to all its customers. The test suite is release specific because each release most probably introduces some extensions that were not present in previous SDK versions and these extensions need to be covered.

Customer projects can start using the test suite at their earliest convenience. Once the program is close to shipping and no changes to the software are planned, it is time to carry out the final verification round.

As shown in Figure 4-4, the test suite produces a report, which should be sent to MSW for final acceptance. The phone can be shipped if no problems are discovered during testing. All possible

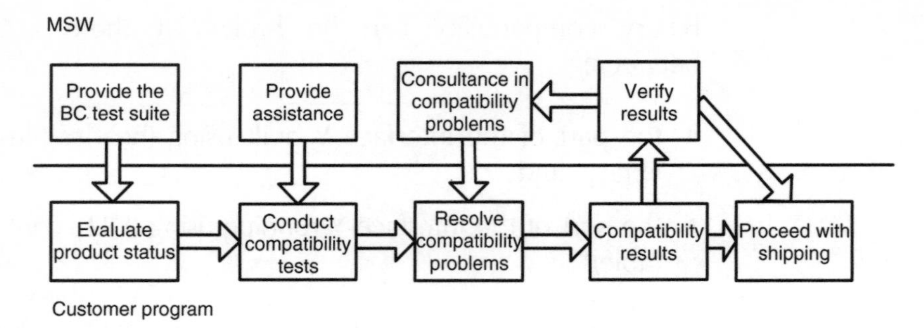

Figure 4-4. Binary compatibility verification process.

breaks need to be corrected and the test suite to be re-run until full binary compatibility can be guaranteed.

4.4.2 The Binary Compatibility Verification Suite

The S60-based phone Binary Compatibility verification suite consists of several tools, applications and steps. The usage of the tools is very straightforward and automatic.

The tools are called:

- SDK Analyser
- Source Analyser
- Binary Analyser
- Application Launcher

The following sections describe what each tool does, as well as how and why it does it.

4.4.2.1 The SDK Analyser

The SDK Analyser identifies changes between two different versions of the S60 platform or between a S60 platform version and the phone full software package. In other words, it requires two source code packages as input and produces a data file containing information on the differences between the two input data sets as shown in Figure 4-5. This report is then used as one of the two inputs to both the source analyser and binary analyser tools.

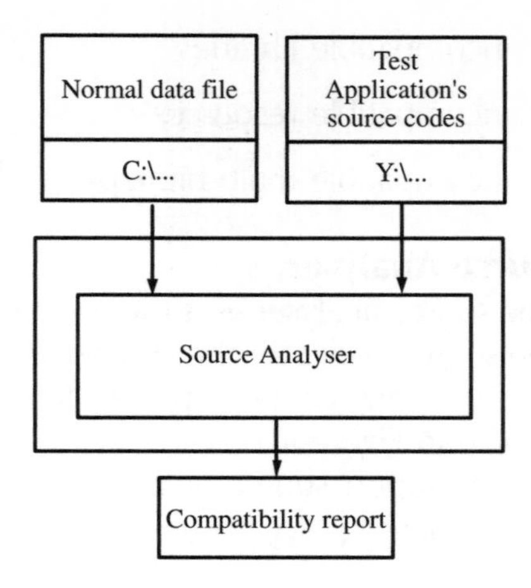

Figure 4-5. SDK Analyser.

The SDK analyser uses certain criteria for its analysis work. The criteria considered are:

- changes in exported public/static function signatures
- changes in function ordinals
- changes in class size
- addition/deletion of exported public/static functions including virtual functions
- changes in the order of virtual functions
- changes in order/size of data members
- changes in constant values
- changes in multi-bitmap definitions
- changes in resource definition

As a result the SDK analyser generates two data filea, a normal one and a compact one. The difference between these two report file is that the normal report includes specific information on the identified differences in definitions of functions, constants, resources and multi-bitmaps definition, whereas the compact file only lists the differences in ordinal numbers. The generated report (a text file) has the following structure:

- incompatible libraries
- incompatible resources
- incompatible multi-bitmaps

4.4.2.2 The Source Analyser

The source analyser uses the normal report as an input together with the source codes for the chosen application. It checks if the application uses any of the exported functions listed in the report file, as shown in Figure 4-6.

The report source analyser generates lists of the following findings, if any:

- changes in the order, number and type of function parameters
- changes in the return type of the function
- changes in constant values

The user can select any file from the report, which enables the following options:

- **Show Details**, which displays a dialogue containing the line numbers where there was a compatibility failure and the reason for it

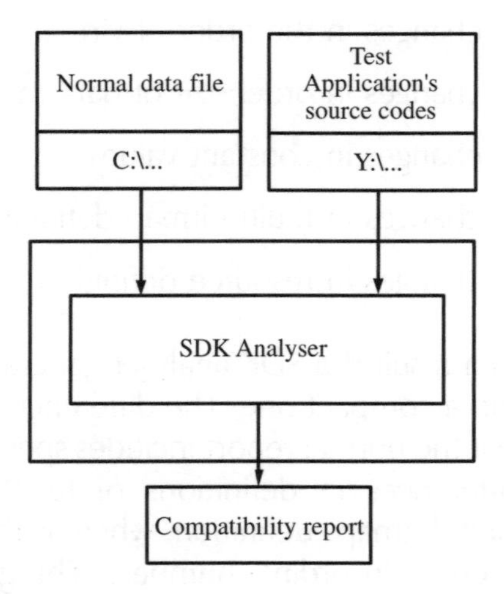

Figure 4-6. Source Analyser.

- **Show Source**; using this portion the user can browse through the source code of the selected file and view each compatibility failure in turn

4.4.2.3 The Binary Analyser

While the source analyser uses the normal report from the SDK analyser, the binary analyser uses the compact report. In addition, the application binaries are also needed to run the binary analyser. Figure 4-7 shows the functioning of the binary analyser, which can be run on either a PC or the target hardware.

The binary analyser reads the header of the ARMI/ARM4/THUMB binary to find the following information:

- the names of the imported DLLs
- the ordinal numbers of the exported functions from each imported DLL

It generates a simple report on the compatibility of the application binaries with the base terminal, the source code which was originally used as an input to the SDK Analyser.

4.4.2.4 The Application Launcher

The application launcher is run on the target hardware only and it performs the following analysis:

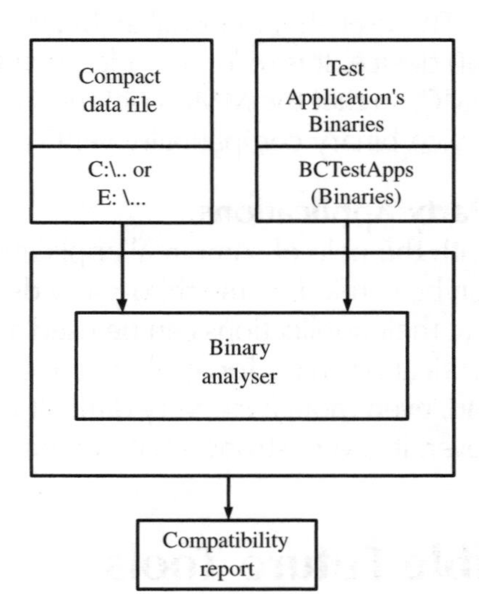

Figure 4-7. Binary Analyser.

- It identifies all imported DLLs.
- It gets the path information.
- It checks for the presence of the resource file path.

In addition, it launches an already installed test application and generates a report if there are missing libraries or resources. The application launcher is a quick and dirty way to check possible binary breaks and therefore it could be used on a regular basis as soon as the target hardware is available. What it does not do is any further analysis if a break is found.

4.4.2.5 Binary Compatibility Applications

The scope of binary compatibility (BC) applications is such as to enable a full coverage of the APIs that need to be checked. These applications are grouped into three main categories:

- the BCApps, which handle the BC testing of specific APIs
- the BCAppLogEngine, which is used by the BCApps to log the test results in a text file (C:\BcAppLog.txt)
- BCAppMain, which is a test driver application enabling automated testing by launching each application from the list of selected BCApps

The user should install all binary compatibility applications in the S60 device. It is recommended that they should be installed into the MMC, so that the MMC contents can be easily reused in some other device binary compatibility verification.

4.4.2.6 Third-Party Applications

With the help of some real applications, the overall logic of the APIs can be verified. Some third-party developers have agreed with MSW that their applications can be used as part of the binary compatibility verification suite. Today, there are over 30 such applications included and, even though it is very difficult to say how much of the SDK they cover, it is very strongly recommended that they should be run.

4.5 Possible Future Tools

Despite how well the tools described above cover the DLLs, the libraries and the resources of the public APIs, there is always room for improvement. For example, what these tools cannot do is to test the logic of how an application is using a certain DLL or function, and therefore if a phone program changes the contents of a DLL in such a way that no visible marks are created, the change in the logic of the DLL can make the application behave strangely.

New tools and procedures, such as the following, may replace the tools introduced in the above sections, either fully or partially. This could make verification faster and more convenient.

4.5.1 DepInfo Tool

A DepInfo tool performs a basic binary comparison between two builds, as shown in Figure 4-8. DepInfo runs on a PC and it compares two builds, copies the resulting differences into the appropriate place and launches Internet Explorer to display the results. The results include differences in the function exports of the DLLs. If the option 'all' is used, then the comparison only covers those DLLs that occur in both builds and the rest are ignored.

4.5.2 Header Checker Tool

The HeaderChecker tool reports possible breaks between two sets of header files. The report can be filtered by using different options such as 'New Exported Functions' or 'Removed Classes'.

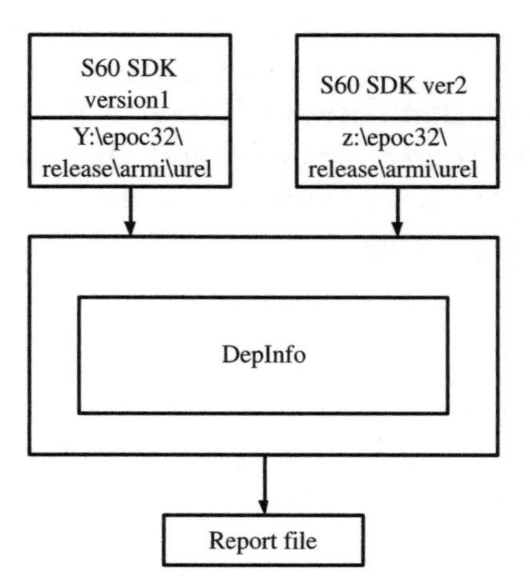

Figure 4-8. DepInfo.

4.5.3 Ordinal Checker

The Ordinal Checker performs a basic binary interface comparison between two builds by comparing exported function ordinals.

These three tools, DepInfo, Header Checker and Ordinal Checker, have all been evaluated and proved to work well. Whether these tools will replace ever the current binary compatibility verification package is still undecided. The phone program needs to consider whether it will get any value-added by using these tools.

A tool that could verify the differences between the logical behaviour of two SDK's does not exist today. Most probably there will not even be one available soon, because, on the basis of current understanding, implementing such tool requires fuzzy logic to be used and therefore this would not only be time-consuming but also money-consuming. It would most probably contain too many faults and therefore the value added by such a tool would be very debatable and not worth implementing.

4.6 Summary

Keeping binary compatibility among different S60-based devices is an absolute requirement if S60 is to really take off. S60 has therefore created a set of tools that verify the device against previously defined BC requirements; these tools have been described in this chapter. These previously defined requirements are basically derived from the set of APIs published on the Forum Nokia web pages so that everyone can utilize them in application development. In addition, the verification process has been explained. Binary compatibility verification uses tools such as the SDK analyser, the binary analyser, the source analyser and the application launcher. Using these tools as suggested provides useful information on a device's compatibility.

Chapter 5: Certificates and Standards

Some of the S60 technologies may need separate certification and some of the features may require third-party licensing plus mandatory third-party acceptance prior to the terminal being shipped. This chapter introduces such features, but, since every Licensee holds its own contracts with these third parties, they should check their liabilities concerning licensing and certification responsibilities as set out in the agreements with the owners of the IPR (Intellectual Property Rights). The sections in this chapter should therefore be read as if they are describing a case, in which the areas listed are understood to be the liabilities of the phone program. Special attention should be paid to Java, which in most cases requires both certification and third-party licensing.

Common to all these separately licensable technologies is that it is very unlikely that any international legal agreement will be reached within a reasonable time. Thus, it may be necessary to involve several lawyers to make all clauses in the contract acceptable to both parties.

S60 Smartphone Quality Assurance Saila Laitinen

The Open Mobile Alliance (OMA) technologies described later in this chapter are based on optional interoperability and verification events for terminal manufacturers, but are highly recommended to each phone program. The S60 Platform introduces some parts of the cellular standards that may be totally new to licensees or have not been taken into use by some operators.

In addition to the above, government and quality certificates are described even where they are not S60 specific because smartphones may have some additional requirements prescribed by different authorities that are not relevant to feature phones.

Some technologies/features may be optional in S60-based phones. However, if a particular feature is used, a license or certificate may be required. Table 5-1 describes all the technologies that either are known to require a specific licence, certification (SW versus HW) or supplementary letter of agreement or are otherwise recommended for official interoperability testing by a third party and with other enabling products such as servers.

5.1 Technology Certificates

Java and Bluetooth are seen as independent technologies requiring licences and/or certification.

5.1.1 Java/TCK

S60 Platform version 3.0 contains among other software the following Java implementation (JSRs):

- CLDC 1.1 (JSR-139)
- MIDP 2.0 (JSR-118)
- JTWI 1.0 (JSR-185)
- Wireless Messaging API (JSR-120)
- Mobile Media API (JSR-135)
- Java APIs for Bluetooth (JSR-82)
- Mobile 3D Graphics API for J2ME (JSR-184)
- PDA Optional Packages for the J2ME Platform (JSR-75)
- Location API for J2ME (JSR-179)
- J2ME Web Services Specification (JSR-172)

TECHNOLOGY	S60-SPECIFIC	HARDWARE	SIDE LETTER	LICENCE	CERTIFICATE	IOP
Java	X			X	X	
Bluetooth	X	X			X	
PC connectivity	X			X		
Predictive text input				X		
Chinese fonts				X		
Chinese dictionary				X		
MIDI engine				X		
Audio & video Decoder release 2.1				X		
Security certificates	X		X	X		
OMA	X					X
MMS						X
WAP						X
IMPS						X
SyncML						X
Content download						X
DRM						X
Client provisioning						X
Government & cellular					X	

Table 5-1. S60 certificates and licences.

- Security and Trust Services API for J2ME (JSR-177)
- Wireless Messaging API 2.0 (JSR-205)
- SIP API for J2ME (JSR-180)
- Scalable 2D Vector Graphics API for J2ME (JSR-226)
- Advanced Multimedia Supplements (JSR-234)
- Nokia UI API

All of these JSRs need to be separately licensed by the Licensee prior to being released as part of the platform.

Symbian tests its own OS release as a software product and the platform does the same for its implementations. If a phone has Java, the phone most probably must execute and pass the Technology Compatibility Kit (TCK) test suite for each JSR. If that is the case, then the Licensee has to license the TCK test suite from Sun Microsystems and self-certification forms have to be sent to Sun.

TCK testing verifies that all Java APIs in the device are correctly implemented. For each API there is TCK and all TCKs need to pass before the JAVA certificate can be obtained. Each TCK includes certain number of test cases. For example, CLDC 1.1 includes 11 500 test cases and MIDP 2.0 over 600 automated cases, which have to be run in the so-called Trusted and Entrusted security domains.

Nearly all test cases are automated and they are run using the JavaTest harness provided with the TCK. Before these tests are run, the TCKs must be installed; in the other words, the testing environment needs to be built.

Running the tests and exercising the necessary troubleshooting with the failed cases requires special know-how on TCK and the JavaTest harness and thus Java programming skill as such does not guarantee that a person is capable of completing these activities. In most cases improperly implemented APIs are not the cause of the problems, but more probably problems in the test environment (e.g. problems in firewalls or poor GPRS connection). The special competencies in TCK testing include, for example, the following: testing proficiency in demanding industrial settings; understanding of PC and mobile hardware; connectivity and compatibility issues; and a thorough understanding and knowledge of TCK testing and its large volume of documentation.

Very often, the extent of the things mentioned above surprises the management of the phone program. Therefore it is recommended that enough time should be reserved for planning and testing the TCK environment and for guaranteeing that the required competencies for running the test and possibly analysing the results are in place and available when needed, as well as that consideration should be given to out-sourcing this part of the program to a professional company fully dedicated to this sort of testing activities with the needed know-how on building the environment, running the tests and carrying out the necessary troubleshooting for failed cases.[1]

5.1.2 Bluetooth

Bluetooth is a wireless connectivity standard. It is becoming more and more popular in mobile devices as well as in office environments. Bluetooth can be used for transferring files and data packages between two devices, for example for printing or when a wireless mouse is used. S60 contains the following profiles in Bluetooth:

- Dial Up Networking Profile (Gateway)
- Fax Profile (Gateway)
- Object Push Profile (Server and Client)
- File Transfer Profile (Server)
- Hands Free Profile (Audio Gateway)
- Headset Profile (Audio Gateway)
- Basic Imaging Profile (Image Push Server and Client)
- Sync Profile (OMA DS)
- Remote SIM Access Profile

A Bluetooth (BT) qualification is needed for all BT products. In order to obtain one, the Licensee needs to be a member of the Bluetooth SIG.

5.1.2.1 BT Certification Areas

BT consists of three layers, one layer each falling into the territory of Nokia, Symbian and the Licensee. These three layers are:

- Physical level, radio part
- Protocol and profile
- Application level profile

A phone program should carry out the physical-level certification, as it is very hardware dependent. The physical-level qualification consists of radio frequency and base band measurements. Typically, BT component suppliers have certified the components. If this is the case, a phone manufacturer only needs to make sure that antenna, clock signal and heat range fulfil the requirements.

The protocol-level certification involves the Licensee, S60 and Symbian. The protocols are implemented in Symbian OS. S60 needs to state that the platform implementation complies with the Symbian implementation, while the Licensee has to make sure that it does not negatively affect the BT functionality when modifying those parts of the S60 platform.

Responsibility for the application-level profile certification belongs to the phone program. The Implementation Conformance Statement (ICS) of profiles in the S60 platform can be used. Some profiles may require implementation in DOS, for example the SIM Access Profile (this applies only to platform versions 2.6 and earlier), while the majority of OBEX-based profiles are implemented in Symbian and the S60 Platform. However, this information should be verified from the related documentation, such as the product specification.

Since Bluetooth devices use radio signals in the Industrial, Scientific and Medical (ISM) band, they must pass the regulatory tests. There are different national and international regulations in this area. Even if it is not illegal, the use of Bluetooth devices that have not passed the tests is not recommended. In some countries it may not even be possible to get type approval for Bluetooth. More information to be found in the following standards: ETS 300-328 and ETS 300-826 European Telecommunications Standard (Europe) and FCC&15 Federal Communications Commission (USA).

A Bluetooth sniffer is a useful tool for testing at the protocol and profile levels. At present, there is some automatic BT test equipment available but unfortunately some of these tools may give different results for the same tests. It is therefore recommended that enough time be reserved for BT certification so that possible interpretations of the differences can be resolved.

The Bluetooth SIG arranges 'unplugged test fests' so as to provide a cost-efficient way to test BT products. In addition, the BT device manufactures get to check the interoperability of their devices with the latest BT devices from other manufacturers. The test fests are arranged every three months.[2]

5.1.3 Other Technology Licences

Other optional technologies requiring a separate licence are, for example, PC Suite, the Midi Engine and the Predictive Text input engine.

The IPR of the PC connectivity involves several parties. Furthermore, there are significant differences between the S60 releases because of the connectivity software dependencies for the version of the Symbian OS that is used. In the S60 Platform Release 2.1 the PC Suite is made by Nokia and is licensed under a separate agreement between Nokia and the Licensee. This PC Suite uses OBEX and SyncML standards. OBEX and SyncML both come from Symbian/Nokia and do not require separate agreements to be in place.

The MiniBAE 1.6 MIDI engine can be obtained from Beatnik Inc. It is necessary if playback of MIDI files is required,or example to enable MIDI ringing tones in the phone.[3]

With predictive text input it is possible to enter text on a mobile phone using just one key press per letter. Predictive text input is a common replaceable component from Tegic Communications Inc. If a terminal has the feature, a licence agreement with the relevant third party is obligatory.[4]

5.1.4 Security Certificates

A supplementary letter of agreement between the S60 Licensees and Nokia is necessary to fulfil the requirements of the following three certificate issuers: Verisign, Baltimore, and Entrust. In practice, it includes Verisign requirements for trademarks as well as Entrust warranty requirements:

- **Verisign:** Verisign delivers infrastructure services that make the Internet and telecommunications networks more reliable and secure.[5]
- **Baltimore:** the Baltimore UniCERT, Certificate Authority (CA), has been extended to produce WAP digital certificates (WTLS certificates). WAP gateways and servers need these certificates to authenticate themselves to mobile device users.[6]
- **Entrust:** Entrust creates software to secure digital identities and information.[7]

Delivery of Nokia's Verisign, Baltimore and Entrust certificates is not included in the S60 Platform OEM delivery as a default. The certificates will be delivered separately to the licensees that have signed the supplementary letter of agreement. Certificates are documented in the S60 product specification.

5.1.5 Universal Serial Bus

Universal Serial Bus (USB) can be used to connect S60-based phones to other devices such as peripherals and computers. USB is optional in S60. Furthermore, certification is not required even if a terminal has USB.

If a phone implements USB, the manufacturer can turn to the USB Implementers Forum (IF) Inc., which is a non-profit corporation founded by the group of companies that developed the Universal Serial Bus specification. The USB-IF Forum facilitates the development of high-quality compatible USB peripherals (devices) and promotes the benefits of USB and the quality of products that have passed compliance testing.

The USB-IF instituted Compliance Program provides reasonable measures of acceptability. Compliance Workshops are held about once a quarter in various locations and typically run for three days. The USB-IF provides special test teams who perform the tests developed for the Compliance Program. Private test sessions are also scheduled for system vendors and peripheral vendors. During these test sessions, the vendors check that their products work well together. Products that pass this level of acceptability are added to the Integrators List and have the right to license the USB-IF Logo. If a product is not on the product list, it does not mean there is anything wrong with the product; instead it proves that any product on the list has passed the tests and is in a way an additional marketing element.[8]

5.1.6 Infrared Connectivity

Infrared certification is fully optional. The mission of the Infrared Data Association (IrDA) is to promote interoperability between IrDA devices. It is the responsibility of each S60 licensee to obtain the needed product compliance in accordance with IrDA specifications. IrDA compliance relies on self-testing by device manufacturers.

IrDA Data protocols consist of a mandatory and optional set of protocols. These protocols are:

- PHY (Physical Signalling Layer)
- IrLAP (Link Access Protocol)
- IrLMP (Link Management Protocol and Information Access Service IAS)

The PHY protocol implementation belongs to the licensees. The following profiles are supported in the S60 platform release 2.0:

- IrDA Stack: IrLAP v1.1, IrLMP v1.1, IrTinyTP v1.1, IrComm v1.0
- Obex: IrObex v1.2
- Digital Camera Connectivity: IrTranP v1.0

The IrReady trademark is optional. The Infrared Data Association has its IrReady program, which defines the minimum set of requirements that will lead to interoperability and an acceptable user experience. The IrReady qualification is then awarded to devices that meet those standards.

The IrReady Program Reference Document, Test Specifications and Profiles will help to properly certificate products. Under the direction of the IrDA Test and Interoperability Committee, IrDA has authorized Interoperability Test Labs to provide testing for devices that wish to have the IrReady trademark. This is part of the certification process and is a requirement for obtaining the intellectual property rights to use the IrDA IrReady trademark.[9]

5.1.7 Multimedia Cards (MMC)

The S60 Platform supports MMC cards. S60 licensees are recommended to contact the Multi Media Card Association (MMCA) but there are no certification requirements. The MMCA develops and regulates open industry standards that define all types of multimedia cards and drives worldwide acceptance of multimedia cards as an industry standard across multiple host platforms and markets. The organization works toward full interchangeability and compatibility (including backward compatibility) between the cards produced by all MMCA members.

MMCA defines the testing procedure for MMC card manufacturers only. There are no specified conformance testing or acceptance requirements for mobile phone manufacturers.

MMCA organizes 'plug fests' for members and non-members. A large array of host and card products compatible with the Multimedia Card standards are brought together in one place. Participating companies have an opportunity to test and verify the interoperability of host platforms and multimedia cards. Test results are always confidential between card manufacturer and host manufacturer. MMCA will not collect or monitor test results. Actions to correct interoperability issues are worked out manufacturer to manufacturer rather than through the MMCA.[10]

5.2 The Open Mobile Alliance (OMA)

The Open Mobile Alliance verification and interoperability sessions are optional for S60 Platform licensees. There are no formal certificates required after successful verification sessions. OMA defines its mission as 'to grow the market for the entire mobile industry by ensuring seamless application interoperability while allowing business to compete through innovation and differentiation'.

It is recommended that all S60 licensees familiarize themselves with OMA IOP policy and process. Nokia MSW tests the S60 platform delivery. Responsibility for the testing of functionalities and interoperability of market-ready terminal products belongs to the S60 licensees even though there are no formal certification requirements. Chapter 9 on Testing as a Tool provides more information on the OMA and on overall interoperability testing.

5.2.1 Process and Principles

Chapter 9 describes the interoperability process, policies and principles of OMA. The interoperability activities in OMA provide for the verification of technologies in products against the technical specification requirements and acceptance of results. The aim is to avoid multiple testing structures and to achieve cost efficiencies. It is intended that the OMA Interoperability Program will evolve as the OMA Enabler Releases evolve. It will be extended and modified to encompass future technologies as determined by OMA. Readers should therefore verify the up-to-date process directly with the OMA.

Interoperability is the key to the success of services based on the standards defined by OMA.

OMA IOP programme includes several methods:

- OMA Test Fest
- Manufacturer bilateral testing
- Testing in a test house

The testing process when the Test Fest approach is used has two phases: Test Fest preparation and Test Fest operations. Test responsible is the administrative group authorized by OMA.

Up-to-date information about current events, their scope and schedule can be found from the OMA web pages at ⟨http://www.openmobilealliance.org/⟩.

To schedule the OMA test fest as part of the phone program, management should take the schedule in Table 5-2 into account.

OMA technologies in the S60 Platform include the following:

- **Multimedia messaging (MMS).** The MMS features of a S60 platform release are tested by Nokia on reference hardware. A Licensee needs to decide how extensive additional MMS IOP testing it plans to be carry out.
- **Wireless Application Protocol (WAP).** The OMA board of directors made a decision that the WAP1.2.1 certification program should be shut down. The OMA conformance and interoperability testing replaces this former WAP certification. OMA also provides a test suite for testing WAP. This test suite is available to all OMA members and it also contains tests that are not applicable to S60-based phones.

ITEM	TIME
Test Fest schedule	–12 weeks (before test fest)
Fest announcement	–8 weeks
Event venue	–8 weeks
Test documentation approval	–8 weeks
Registration opens	–8 weeks
Registration closes	–2 weeks
Deadline for submitting test material	–1 week
On-site information sent to participants	–1 week
Detailed on-site information available to participants	–
Test Fest testing	–
Notification of Test Fest results	+1 week
Publishing Test Fest information on OMA website	+1 week (after Test Fest)

Table 5-2. OMA Test Fest preparation guidelines.

The S60 Platform release 2.0 contains an XHTML mobile browser, which enables users to access services related to the WAP 2.0 specification. The S60 Platform WAP Implementation Conformance Statement will include the Static Conformance Requirements (SCR) from all of the June2000 WAP specifications.

The Instant Messaging and Presence Server (IMPS) is also covered by OMA and therefore does not require any additional license or certificate. OMA has arranged optional interoperabilities (IOPs) since November 2002.

Registration usually ends two weeks before an event. Members can register for these events on the OMA web page.

SyncML in S60 supports the following synchronization types:

- two-way synchronization
- slow synchronization

The protocols supported are:

- HTTP 1.1
- WSP (WAP 1.2.1)
- SSL/TSL can be used for security
- WTLS is not supported
- VCalendar v1.0
- VCard v2.1

SyncML testing is optional. The SyncML client, which is a core component of Symbian OS, is an open standard that uses a common language for communications between devices, applications and networks. The SyncML open standard ensures a consistent set of data that is always available on any device or application at any time. SyncML Device Management (SyncML DM) enables OTA administration of devices and applications, simplifying configuration, updates and support.[11]

When the SyncML initiative was assimilated into OMA, it was accommodated into the OMA interoperability (IOP) process. At the same time the SyncML acceptance procedure was terminated. The purpose of the OMA fest and bilateral IOP testing is to improve the interoperability of the specification by testing the interoperability of separate devices. Passing the tests successfully at an OMA fest no longer entitles a product to be listed, as was the case when the SyncML initiative was used.

Digital Rights Management (DRM) enables content providers to associate certain rights with the content objects that define how the content can be used. The OMA DRM 2.0 standard, which is supported in S60 version 3.0, provides various levels of DRM methods for mobile content delivery: forward lock, combined delivery, and separate delivery. Forward lock prevents an end user from sending DRM-protected content to other end users.

The DRM implementation on the S60 Platform is ready for CMLA, but it is not certified, since only devices can be certified by the Content Management License Administrator (CMLA).

5.3 Cellular Standards and Operators

Not only cellular standards, but also operators set additional requirements for terminal manufacturers both in technology and in usability.

The European Telecommunications Standard Institute (ETSI) is a non-profit-making organization whose mission is to produce telecommunications standards. The GSM standard requirements can be obtained from ETSI. The S60 Platform supports the ETSI 3GPP TS 51.010-1 V5.1.0 test specification. Support in this case unfortunately does not guarantee that a Licensee's S60-based phone will also fulfil those requirements; instead it only guarantees that the platform does not contain anything that prevents the requirements from being fulfilled. This is natural since the Licensee has their own implementation for the telephony parts of the phone as well as its own RF solution. Most of the phones aimed at the GSM markets need to pass some version of the ETSI 3GPP tests. In most cases it is the operator who defines the version that needs to be passed for each phone.[12]

Code Division Multiple Access (CDMA) phone programs should plan CDMA features carefully and reserve enough time for operator acceptance. This is because of the nature of the CDMA standard, which concentrates fully on defining the air interface in which, at the same time, many network interfaces are proprietary. Each operator has to certain extent a unique solution for the composition of the network and the major operators may have different service platform requirements. Whereas some carriers support standard OMA implementations, others use non-OMA deployment. As a result of the fragmentation in the CDMA world, there may be a need for operator-specific software versions. In fact, there are more operator-specific requirements than generic CDMA requirements.

From a licensee point of view, one of the advantages of the S60 platform is that the majority of difficult operator-specific requirements are already implemented in the platform software. However the IOP, certification and acceptance have to be done for the each terminal product separately.[13]

5.3.1 Government and Quality Certificates

Government-based certificates are typically common to all phones irrespective of their feature set. This section provides a look at what kind of additional requirements based on different governmental legislations terminal manufacturers may have to face when trying to sell the product. Several authorities and organizations may have some 'certification by similarity' procedures for 'copy' products, but the licensees should not assume that the terminals automatically fulfil these similarity definitions, which vary between organizations. Furthermore, the Licensee should treat this list as incomplete because the legislation is ever-evolving and ever-changing. Thus, every phone program should acknowledge its responsibility in figuring out the prevailing requirements for the phones to be the shipped.

5.3.1.1 Mandatory

EU Directive RTTE 1999/5/EC is mandatory for obtaining the CE mark. The licensees who wish to self-certify products can use the method described in Annex V, which is one of the alternative ways to demonstrate compliance with the requirements of the Radio and Telecommunications Terminal Equipment Directive (R&TTE 99/5/EC). The route using Annex V is known as 'Full Quality Assurance Approval', because it is based on the evaluation of the whole quality system. In practice, this means that a vendor can self-certify new GSM products quickly by testing them in their own or an external accredited test laboratory and send the results to the notified body.[14]

European Commission Automotive Directive 95/54/EC describes how to obtain the e-mark that is mandatory for devices that are to be connected to the power supply of a vehicle. Some testing is also required for this.

The Restriction of Hazardous Substances (RoHS) is an EU directive. It may be extended to other geographical areas some day as it has an environmental aspect. Products containing lead, mercury, cadmium, hexavalent chromium, polybrominated biphenyls and polybrominated diphenyl ether must not be sold in the EU after 1 July 2006. Producers will be responsible for taking back and recycling electrical and electronic equipment. This will provide incentives to design electrical and electronic equipment in an environmentally more efficient way that takes waste management aspects fully into account. Consumers will be able to return their equipment free of charge. There may be some restrictions on the materials used in terminals.[15]

Waste Electrical and Electronic Equipment (WEEE) is a draft EU directive. It instructs the user to dispose of WEEE separately from other waste.

Local Approvals for Games

Since the S60 Platform enables licensees to include a broad-spectrum of games in the terminals, it is recommended that licensees verify the local requirements for game approvals from the relevant authorities. The latter include:

- Interactive Software Federation of Europe (ISFE)
- Entertainment Software Rating Board in the USA (ESRB)

Prior approval for games is required currently in approximately 15 countries in the Europe and Africa region. After ESRB approval is granted, a sticker and registration number with an age limit is received for sales packages.

The Federal Communications Commission is a mandatory certification required in the USA. The Federal Communications Commission (FCC) is an independent United States government agency. The FCC is charged with regulating interstate and international communications by radio, television, wire, satellite and cable. Required actions are testing against FCC requirements paperwork and type approvals.[16]

The PCS 1900 Type Certification Review Board is a certification that can be required by operators and US market. The purpose of the PTCRB is to provide the framework within which GSM Mobile Equipment Type Certification can take place for members of the PTCRB. This includes, but is not limited to, determination of the test specifications and methods to implement the Type Certification process for GSM Mobile Equipment. Required actions are testing against PTCRB requirements, paperwork and type approvals.[17]

China approvals and certifications can in some cases set additional requirements for terminal programs trying to get into Chinese markets. They may require testing against local requirements, paperwork and type approval.[18]

Local approvals and certifications may be needed for areas where EU, USA, Canada, China approvals are not enough. As an example these could be required by:

- Australian Communications Authority
- Canada (see ⟨http://www.crtc.gc.ca⟩)
- Japan (see ⟨http://www.soumu.go.jp/⟩)

5.3.2 Optional

Global Certification Forum (GCF)

Participation and membership in the Global Certification Forum is voluntary. GCF is a partnership between operators and terminal manufacturers. It provides an independent program to ensure global interoperability of 2G and 3G mobile terminals. In addition, other parties involved in terminal development, including test houses and testing equipment manufacturers, may participate as observers.

Benefits of membership are as follows:

- restricted PRDs
- meetings and meeting documents
- 3G activities
- test cases
- five- and ten-day rule document approval process
- field trial (FTQ) documents
- terminal information and documents
- membership database
- other GCF internal documents

The current membership includes over 140 network operators worldwide, 30 leading terminal manufacturers and 48 test equipment manufacturers, test laboratories and other observers.[19]

Cellular Telecom Industry Association (CTIA)

CTIA is an optional certification in the US market. Involvement in the discussion forum and in testing activities is required in the CTIA. When a phone passes certification, the manufacturer has the right to exhibit the CTIA Certification Seal on the phone and its packaging and to use the CTIA Seal in its advertising.

Certification applies for the following technology platforms in North America:

- CDMA
- GSM
- TDMA
- analogue

CTIA-certified products must pass a rigorous three-part technical evaluation. All test plans can be downloaded from the CTIA web pages:

- Part 1 tests the product's conformance to the wireless industry's technology platform standards. These tests, which are conducted by a CTIA Authorized Testing Laboratory, are defined in the test plans available from CTIA.
- Part 2, which is also conducted by a CTIA Authorized Testing Laboratory, tests a product's over-the-air performance and is defined in a test plan.
- Part 3 of this evaluation is the FCC Type Acceptance Testing. An FCC authorized testing laboratory conducts this testing.

CTIA Certified products are required to include information for consumers about important health and safety information related to the use of wireless products.[20] This information includes:

- driver safety information
- consumer safety information

- audio accessibility
- hands-free capability

Entertainment & Leisure Software Publishers Association Europe (ELSPA)

ELSPA was founded to establish a specific and collective identity for the British computer and video games industry.[21]

ELSPA addresses the following issues:

- industry promotion
- sales charts and reports
- conferences and seminars
- anti-piracy enforcement
- reviewing proposed legislation
- content ratings
- research reports
- careers promotion
- ISO Standardization

International Organization for Standardization (ISO)

ISO is the world's largest developer of standards. It is the world leader in providing widely accepted quality-related standards to the market. Some of these quality standards, if followed, can provide company-wide certification and therefore can be applied to S60-based terminals as well as to any product a company produces. The licensee's task is to determine which standards are applicable to their products. The following are some of the possible certificates:[22]

- The ISO/IEC 17025 Laboratory Quality Standard enables laboratories to issue accredited test reports with traceability to national and international measurements standards.
- ISO 9000 Certification requires quality manuals and internal audits, as well as that management reviews take place as part of every project.
- The ISO/TS 16949:2002 certificate is an international technical specification for the car industry, specifying the quality-

management system requirements for development, production, assembly and service of automotive-related products.

5.4 Summary

S60 includes numerous features that need either separate certification or a license from a third party. Such a feature is, for example, Java Specification Requests (JSRs). In addition, there are features and technologies that do not require certification, but for which it is highly recommended that certain interoperability verifications be carried out in order to confirm correct functionality. This chapter has introduced many of the requirements of the certification and interoperability verification processes. The list of technologies covered is by no means a complete one as the actual requirements are as stated in the contracts between the licensing parties and not, therefore, necessarily as described here.

Chapter 6: What Quality Means

People define quality in many ways. A high-quality product to one person can mean unacceptable quality to somebody else. Very often people identify quality with a lack of defects in a product or service. In addition, what the term 'defect' means different things to different people. Equally, the product feature set is an important aspect of the definition of a high-quality product, as well as the timing of when these features are made available. This chapter provides a look at quality from different perspectives such as quality culture, quality standards and quality in a software product.

The American National Standards Institute (ANSI) and the American Society for Quality (ASQ) define quality as follows:

> the totality of features and characteristics of a product or service that bears on its ability to satisfy given needs.

All self-respecting companies agree that the main reason to pursue quality is to satisfy customers. This is also called fitness for use. In highly competitive markets, those who succeed do not only meet customer expectations but generally exceed them. Thus, one of the

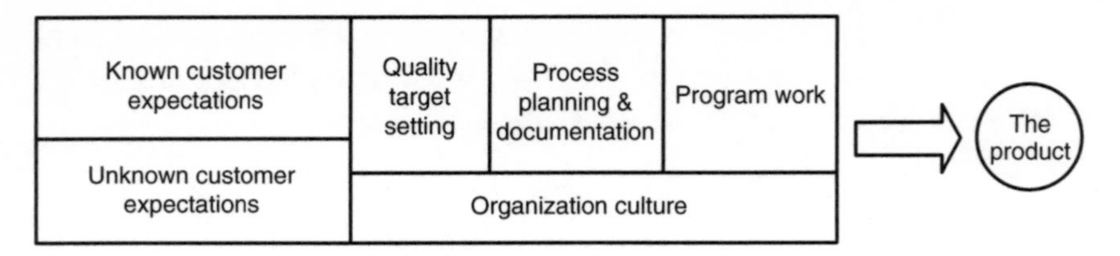

Figure 6-1. Product quality diagram.

most popular definitions of quality is meeting or exceeding customer expectations.[1]

Quality expectations and the challenge in fulfilling them are shown in Figure 6-1.

Organizational culture has a definite impact on the success of a company in the markets that it serves. Kimberley Kingsley wrote:[2]

A relational based quality is grounded in organizational integrity and offers a strategy for organizations to realize their full potential. No matter is it a physical product or a service the quality is either built in it or not by the individual employees.

6.1 Quality Culture

Quality culture indicates, as mentioned above, the organizational integrity in failing, meeting or exceeding customer expectations. An organization, whether it is small or big, if it is led properly, can have a common goal that is well-communicated widely understood. This is necessary if business success is to be achieved. According to Kimberley Kingsley's paper,[2] the five underlying principles of relationship-based quality offer clarity of purpose and a path by which any organization can discover its potential:

1. *Adopt a global perspective*; the birth of Internet has shrunk the world and made accessing anyone, anytime, anywhere very simple and easy. If this is understood correctly, the organization can better interpret what customers expect and when they expect it.
2. *Invoke organizational integrity*; people tend to perceive the motivation and character of other people. In the other words, teams or groups of people have been able to reach tremendous targets mainly because they were so focused and motivated.

3. *Apply core values*; the company values need to be defined, demonstrated and disseminated. Subsequently, it may also need to be reflected and modified.
4. *Inspire leadership*; relationship-based quality inspires leadership at every level by bringing out the leader in each person.
5. *Embrace transformation*; once values and targets have been embedded into the organization, people start to work together towards the one target. Their working preferences automatically become synchronized.

Before the industrial revolution, craftspeople surely understood that customers expect high-quality products. At some point in history, some companies tried to influence their markets and customers by manufacturing goods to meet customer needs, which either existed or did not exist at all. This still happens, but in a more effective way, and a real customer-oriented approach has developed in all world-class companies irrespective of the industry concerned.

How can the understanding of the customer arise? James R. Evans and James W. Dean Jr in their book *Total Quality*[1] introduce the following principles for encouraging better understanding of the ultimate expectations of customers:

- **Collect customer information.** The usage of customer satisfaction surveys has increased significantly over the past decade. It seems nowadays that you cannot visit a restaurant without being given a customer feedback form and asked to fill it in and return it to reception. This is easy to understand since one of the most popular ways to collect customer information is customer surveys. Who else can provide more truthful feedback on a product than the person using it? If a survey is well designed, it can provide vital information on existing customer penchants for a specific product group. The Japanese automobile industry has taken this even further. Teams of automobile designers visit people's homes and observe how they live in order to anticipate their automotive needs. For example, one executive vice-president of Honda has said: 'We should not try to sell things just because the market is there, but rather we should seek to create a new market by accurately understanding the potential needs of customers and society'.

- **Disseminate customer information.** Too often an engineer designs his or her product for another engineer even though the potential customer may be in a totally different profession. The 'bigger' the product, the more distance there is between the individual employee of the manufacturer and the customer. This creates a potential gap in employee understanding of the customer and it is why it is vital that the information received through surveys is well communicated to all employees. This information must be translated into the features of the organization's products and services.
- **Use customer information.** After the customer satisfaction information has been communicated to everyone, the following question needs to be answered by all employees: how can we as a company improve customer satisfaction?
- **Manage customer relationships.** One of the most critical roles in any company is that which has responsibility for the customer interface. Customer contact employees are the people whose main responsibilities bring them into regular contact with customers – in person, by telephone or through other means. Despite all efforts to satisfy customers, every business experiences unhappy customers. That is why customer contact people should also be well trained to receive customer complaints.

6.2 Quality Standards

As quality has become a major issue in business, various organizations have published standards and procedures about the topic. Probably the most well known quality standard, the ISO 9000-family, is introduced in this chapter. In addition, the Six Sigma method is briefly introduced in Section 6.2.2.

6.2.1 ISO 9000

The International Standardization Organization (ISO) was founded in 1946. It is made up of representatives from the national standardization bodies of 91 nations. The agency adopted a series of written standards on quality in 1987. The outcomes were revised in 1994.

The ISO 9000 standard family defines *quality system standards* and its structure is shown and in detail in Table 6.1.

The ISO standards were originally intended to be advisory and to be used for two-party contractual situations between a customer

ISO STANDARD NUMBER	FOCUS
9000	Principal concept of quality assurance
	Objectives of quality
	Responsibilities for quality
	Stakeholder expectations
	The concept of process
	The role of processes
	The roles of documentation and training in support
	How to apply different standards
9001	A model for QA in firms that design, develop, produce, install and service products
9002	A model to firms engaged only in production and installation
9003	Applies to firms engaged only in final inspection and test
9004	Guides the development and implementation of a quality system
	Examines each of the elements of the quality system in detail
	Can be used for internal auditing purposes

Table 6-1. ISO standards for quality assurance.

and supplier as well as for internal auditing. However, the standards quickly evolved into criteria for companies that wished to be certified by a third party on their quality management procedure.

6.2.2 Six Sigma

Six Sigma is a widely known and accepted statistical measure typically used in measuring the variability of a given process. The American Society for Quality Six Sigma states the following:[3]

It could measure for example the number of defects in a subassembly or in a service environment; it could quantify delays in end of month reconciliation procedures. According to leading estimates most companies today are operating at levels of around four sigma, or approximately 6000 defects per million. When a company has achieved a Six Sigma rate of improvement, it has reduced defects to 3.4 per million, which is virtually defect free performance.

Six Sigma is a committed management approach to quantifiably solve problems and optimise critical processes. Adapting and applying the six sigma method can lead to dramatically improved business performance and bottom-line profitability.

As an example, Samsung Electronics Co. Ltd. published a paper in 2004 that describes how to combine quality and speed to market.[4] According to the paper, the Six Sigma's Define, Measure, Analyze, Improve and Control (DMAIC) method was adopted by the company in 2000 to prevent anticipated problems and to gather feedback data relating to mass production. As a result of the experiment, Samsung Electronics Co. Ltd confirmed that Design for Six Sigma (DFSS) was a powerful method for designing new products and procedures.

6.3 Quality in a Product

The customer, of course, always has certain expectations about the product he or she has purchased. If these expectations are not fulfilled, it is very probable that the customer will choose some other manufacturer next time. Earning back customer respect and trust costs money and time and therefore, once the customer has chosen a certain manufacturer or service provider, this company will most probably try to do everything to please the customer with its product.

As already stated in the beginning of this chapter, quality has very many meanings and dimensions. Probably the most widely accepted depiction of high quality is a lack of defects in the most frequently used functionalities, combined with price compliance. In addition, the feature set and timing, in other words when the product hits the market, are two dimensions of quality.

CASE STUDY

This case study explains how customer expectations have a direct link to customer satisfaction and product quality: Person A and person B are both considering buying a new car. Both want to invest only in a high-quality car. Person A is ready to invest lots of money on it and chooses a top model of a well known make of car. Person B has less money to invest and chooses a less prestigious make of a car. Both cars have a cruise functionality, which holds the speed without a need to keep the foot on the accelerator pedal. Person A's cruise control is behaving in an inconvenient way by increasing the speed whenever the car travels down hill. Person B's cruise control is not working at all. However, Person A may be more disappointed than person B. This is because person A expected to get a perfectly functioning feature, which was important to him. Person B instead thinks that the quality resides more in the low price and therefore accommodated easily to the deficiency of some features.

Different aspects of quality apply to manufacturing and to services. The following chapters define these two types of product and the quality in them.

6.3.1 Quality in Manufacturing

To most of us, a Mercedes Benz probably represents a high-quality automobile. For some of us this can be fairly difficult to justify. However, the car's reputation is such as to confirm, even for someone who has never driven a Mercedes Benz, that it represents a fine-quality car. Even though different consumer groups put different weights on different dimensions of quality, they all share certain common elements. Below are some of the quality dimensions that are common to manufactured products:

- **Product feature set** defines aspects of functionalities that the end user could find beneficial. Features in most cases are the number one reason behind the ultimate decision to purchase. For example, in the mobile phone business the features could include digital camera, WLan, instant messaging and a certain set of connectivity features. The manufacturer naturally tries to choose the feature set to be as attractive as possible to attract the widest potential customer group.
- **Product performance** defines how the chosen feature set works in practice in real life. In the software world this has at least three dimensions, the basic functionality of an independent feature, the overall capability of the whole product, which essentially means parallel usage of the features, and the reliability of the product over the long term. More on the performance issues can be found in Chapter 9.
- **Usability** is one of the most important quality dimensions and unfortunately it sometimes seems as though it has been overlooked in the mobile phone business. Other words for usability are conformance and fitness for purpose. The importance of a product's usability will dramatically increase for smart phones as it is unreasonable to expect the mass of the potential phone buyers to be able to figure out how the technicalities of the phone works and how to get certain technologies into use. The complexities of the newest technologies should be hidden from the end user so that usage of them is simple, convenient and easy. It

is not enough simply to have such technologies; they are only worth using if how to use them can be easily discovered.

- **Serviceability** means how easy it is for the customer to have the product repaired and how long such repair takes. The customer does not get a good impression if he or she has bought a product that is meant to be used in daily basis and after several weeks of use no longer works. The customer takes it to the merchandiser, who then says that the product needs to be sent to the manufacturer and it may take four to six weeks to have it fixed. No matter how appealing the product is and how well it worked initially, the chance of the customer choosing other manufacturer next time has increased.
- **Look and feel** is also an important thing in certain commodity groups such as mobile phones. When phones were first widely available, most value was put on the aesthetics of the hardware, but nowadays the software user interface plays an important role too. A colour display is about to become standard and more attention is to be paid to the attractiveness of the application icons.

6.3.2 Quality in Service

It is said that tourism will be one of the most rapidly increasing industries in the whole world in the next few years. People's ever-evolving mobility has brought in many service providers around the world and will continue to do so. The competition over the customer base will be tough. Who out of many service providers will be successful and who will fail? The world is about to shrink even more and an increasing number of people from different parts of the world will have the chance to travel abroad and broaden their understanding of the foreign countries around us. Tourism is directly linked to services and understanding the importance of a high-quality service being available is vital for service providers all over the world. High-quality service is very often seen to be linked to certain countries and nations. For example, many Asian countries are known for their excellence in service; they have somehow managed to build this into their culture and genes. How do they do that? Is there something other countries could learn from them? Most probably, yes. The following list of quality dimensions in service businesses is common to all service-based products:

- **Timeliness**: is the service that is promised really delivered on time? How much time must the customer wait? People have a right to expect that, if something has been agreed with the service provider, the contents and schedule will hold at least unless they are otherwise informed before the expected delivery time.
- **Exactness**: is the service being performed right the first time? Are all items in an order in place or is something missing?
- **Accessibility to the service** is how easy the service is to get. Nowadays almost anything can be found on the Internet, which has been taken into use as a marketing channel by almost all service providers in the world. However, some older people may still not have access or a willingness to log on to the Internet and therefore prefer paper advertisements instead. Therefore a company whose main business is to provide, for example, cleaning services to the elderly, will most probably still prefer old-fashioned leaflets to be sent directly to the target customer group and use that as a primary advertising channel.
- **Behaviour:** every time a human interaction is involved in the service, the quality of behaviour plays a vital role. Friendliness, empathy and an eye to seeing what the customer wants are more valuable than anything else. However, friendliness can mean different things in different cultures and therefore very many service providers choose to have local employees to take care of local customers.

6.3.3 Getting Better Quality

Quality either is inbuilt or it is missing from a company. As mentioned in Section 6.1, the key element is what kind of quality culture a company has and whether it is well communicated to the entire organization, because, after all, it is the individual employees who hold the power to make the difference in product quality.

Culture has been defined in different ways at different times. Throughout the history of the changing notions of culture, it is apparent that anthropologists have long questioned the discreteness and boundedness of culture as something that can be fixed to a particular group located in space and time.

Globalization, information availability and transnationalism are forcing us to re-think the concept of culture. Cultural differences have started to vanish as information availability became so efficient with the help of the Internet and international travel. Accommodating different consumer habits into one's own life is more likely to happen than ever before.

Quality culture, like any other culture, is evolving over time. *Culture* is very often mixed up with geographical areas and reference is made to the culture of Asia versus European culture. After all, one thing is common to all cultures and that is that a group of people align their behaviour with a particular culture, that is, a set of beliefs, penchants and values shared by a group of people. For example, an average high-school culture includes fashions in clothing, music and hobbies, as well as a special language that is used by most of the pupils in the school.

Company culture is used as a tool to build a quality procedure, which includes the target, processes, tools and people. Target means that in a private company nothing should be done without a financial reason. The quality procedure target could, for example, be to reduce the number of defects found by customers in the product by 30% starting from the beginning of next year. The process is a set of steps, states and actions, which describe how the target is met. Tools means a list of supporting tools that can be used to make transitions take place faster and be more error free. People then means individual responsibility for following the procedure and making sure that all steps are done correctly and in time.

A company culture is naturally influenced by the organization's structure. Some companies organize their teams in functional way, so that each function is carried out by a team specific to the task. The positive thing about functional structures is that each task is performed by people who are specialists in the area of the task and therefore fewer errors are likely to occur. The disdavantage is that it may separate the people carrying out the task from customers, which leads to a situation where customer understanding is at risk. This, in turn, may increase the possibility of misunderstanding customer expectations and to a decrease in quality.

No matter what kind of organizational culture is chosen, the following things should be recognized in order to create a quality culture within the company:

- The focus must be kept on quality processes. Many companies have their own defined and documented quality strategy. Personnel should always be very well trained in this strategy.

- Customers must be widely understood. If bringing an external customer to all employees is impractical, at least the internal customers need to be identified for each work task. Enabling people to 'walk a mile' in the customers' shoes often pays back by improving fitness-for-purpose.
- The use of steering groups can also help in identifying immediate needs for resources, tools or processes.

6.4 Quality in the S60 Platform and S60-based Phones

The S60 platform has gone through extensive usability tests and represents, therefore, one of the best smartphone user interfaces in the world. This section gives some hints at a high level for understanding the tools that should be used when making a high quality S60 based phone. As explained in chapter 3 on release management and baseline selection, a phone program has access to very early platform releases, the stability and overall quality of which can be debatable. This is due to MSW wanting to provide open communication and to enable customers to start their own activities as soon as they want, in other words when the customers find it more advantageous than disadvantageous to start making their own implementation or plans based on the platform version they have in their hands.

6.4.1 Choosing the Process

The overall quality of the S60 based phone is influenced by the process on which the project is based. As the chosen process can have a tremendous impact on a program's success, it should be chosen carefully. One should not forget to tailor the chosen process appropriately so that they fit the organization and the product, because the published processes are always just a base. At best, a process truly serves the whole organization when it meets the given quality targets within schedule and within budget.

Some most widely known development processes are analyzed in the following sections.

6.4.2 The Waterfall Process

The waterfall process, shown in Figure 6-2, is probably the most well known software development process in the world. It has widely

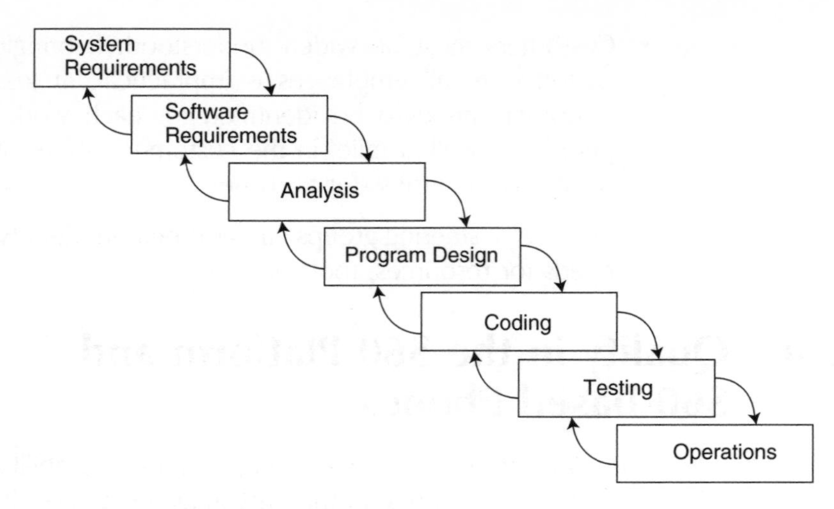

Figure 6-2. Waterfall process model.

used for decades, but its popularity has started to decrease as software size and complexity has increased. Based on Graig Larman and Victor R. Basili's paper in *IEEE Computer Science*,[5] Winstow Royce's waterfall model has been misinterpreted many times, because originally Royce recommended that the process should be followed twice. However, the basic elements and activities remain in almost every software project even if the frequency and swiftness nowadays have to accommodate the requirements set by the software industry.

The absolute strength of the waterfall method is the clear importance of different activities. In addition, as the figure indicates, it allows a step back to the previous activity to be taken. On the other hand, following the waterfall process slavishly most probably puts the overall schedule at risk as the process tends to be very heavy and awkward.

6.4.3 The Incremental Process

The incremental process, which is shown in Figure 6-3, generally provides more confidence about quality as the idea of it is to make the final product in small pieces and bring each piece or feature to maturity before any further development is undertaken.

The advantage of the incremental process is that, when the product ships, one can be sure that the functionality is of good

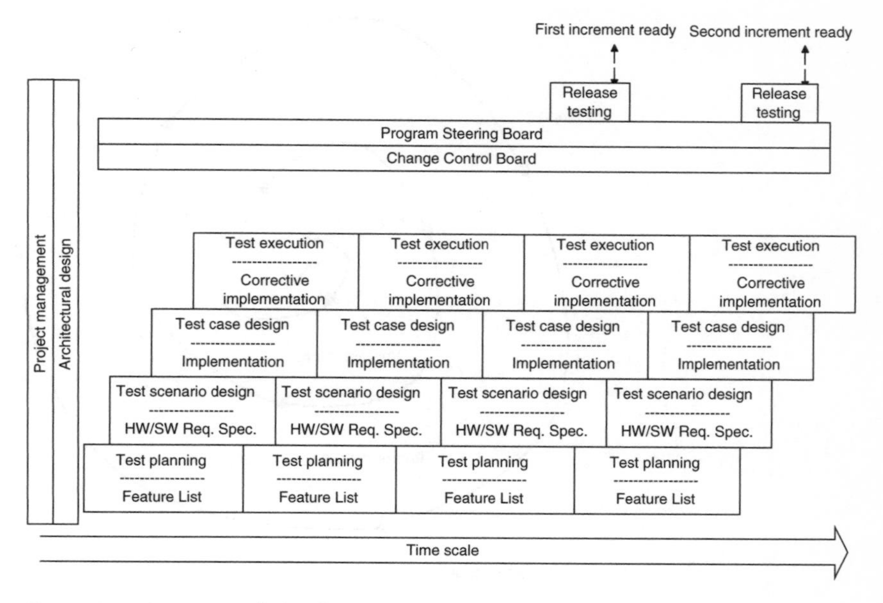

Figure 6-3. Incremental development process.

quality. There is, of course, a trade-off in that some functionality may be completely.

This process brings most value-added rather late of the program when there is no longer a place for sudden drops in stability.

6.4.4 Agile Software Development

The agile method is particularly appropriate when new innovative prototypes are created. The Agile Alliance ⟨http://www.agilemanifesto.org/⟩ has published a manfesto that includes valuing the following four principles:

1. Individuals and interactions over processes and tools
2. Working software over comprehensive documentation
3. Customer collaboration
4. Responding to change over following the plan

Agile process activities are shown in Figure 6-4.

In brief, the agile process allows lots of flexibility so that changes in activities and the focus of the project can be accommodated as

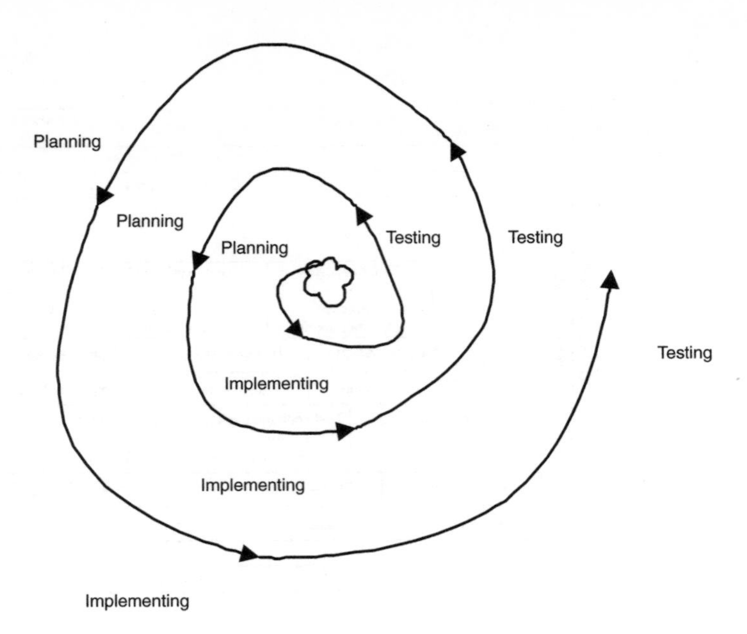

Figure 6-4. The agile process model.

the project moves towards maturity. A drawback of this is that it sets additional requirements on the project management, which needs to keep up to date with status of the project, its sub-systems and completion of components. The agile method was invented mainly for the following reasons;

- The time given to make a product has decreased.
- The given requirements change over time.
- Quality criteria contradict many aspects of 'old-fashioned' software products.
- It is sometimes crucial to be the first to get the product out to the market.

6.4.5 Concurrent Engineering

In product development programs nowadays the chosen process should adopt very easily to the what is called concurrent engineering. This applies especially to smartphone programs. Concurrent engineering means that all key activities are contributing to the project at the same time. This sounds a bit chaotic but is in fact the only

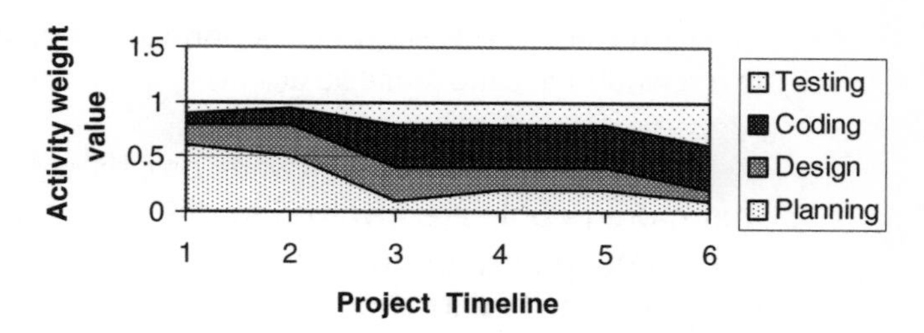

Figure 6-5. Weight values of concurrent engineering activities.

process worth following in multi-supplier projects of relatively large size.

In concurrent engineering, activities within the project are given weight values, as shown in Figure 6-5.

Making a S60 phone sets additional requirements on process flexibility, because activities should accommodate quickly to the sometimes unpredictably changing maturity of different components and sub-systems. Unpredictability can be caused by, for example, the possible internal dependencies of the sub-systems and components, which had not been recognized when the decision on making a certain type of modification or extension to the phone took place.

Very often, none of the approaches described above are used as such within a project. As mentioned in chapter 3, the S60 platform development follows a process combination of several approaches.

6.4.6 Other Things to Consider

Let us assume that the chosen method seems to fit the organization just fine. This section introduces some other aspects that may need extra consideration in each S60-based phone program as they may have an effect on the achievements of the program.

Many companies tend to sub-contract parts of the production to other companies and there are different practices in how sub-contracting is performed. As Evans and Dean write in their book:[1]

> Although the principles of the Customer-Supplier-Relationships (CSRs) are the same with suppliers as they are with the customers, the practises are somewhat different. In general the fundamental practises for dealing with suppliers are (1) to base purchasing decisions on quality as well as cost, (2) to reduce the number of suppliers, (3) to establish long-term contracts, (4)

to measure and certify suppliers' performance, and (5) to develop cooperative relationships and strategic alliances.

The sizes of terminal software programs have grown and more and more people are involved in different activities in the program. The same principles to some extent could be applied within a company. Each entity should have a clear understanding of who their customers are, what do these customers expect from them, when do they expect it and what are their quality expectations concerning the sub-product that entity is responsible for. The sub-product can be part of documentation, part of the system, part of service or anything the program organization needs that forms one link of the full chain.

To sum up this chapter into a couple of sentences: Quality has many dimensions and the manufacturer has a direct impact on the quality of the product. Manufacturers need to understand, agree and communicate the quality targets based on customer expectations to all personnel in the program. The correct process also needs to be chosen and modified to fit into the organization, including possible sub-contractors and suppliers.

6.5 Summary

The quality concept has become a fashionable keyword in every product around the world. However, quality can mean very different things to different people. Often it is linked to the word 'faultless', but a problem arises when people are asked to define a fault in a product. According to current knowledge and understanding, the word quality has multi-dimensional meanings and therefore quality can be measured in numerous ways. Other possible ways to describe product quality are, for example, time to market, feature richness and usability, just to mention a few. The quality concept in service businesses needs a different description. Timeliness, availability, behaviour and accuracy are maybe the most common words to indicate quality in service. There are many different quality standards in the world today and, with the help of these standards, both the consumer and the product or service provider can understand better what the quality means.

Chapter 7: Stumbling Blocks

The term 'quality', together with ways of improving it, was introduced in the previous chapter. In this chapter, the focus is on things that normally prevent manufacturers fom obtaining good product quality in terms of both time and errors. First, the focus is kept at a more general level, but later sections address in detail the potential stumbling blocks that may threaten the success of a S60-based phone program.

7.1 Stumbling Blocks General to All Projects

In any product program, despite of whether it is a software project, an embedded project or a pure hardware project, there are certain stumbling blocks that can cause increasees in the schedule and/or the budget.

These stumbling blocks can vary quite much depending on:

- the nature of the product
- the company manufacturing it
- possible suppliers in the supplier chain

The nature of the product itself can make it more liable to certain types of risk. For example, a program making an expensive and critical part for a space shuttle most probably focuses its risk analysis on defects, which can cause danger for the pilots. Therefore, the National Aeronautics and Space Administration (NASA) has created its own tool for risk analysis. This tool is probably the most highly developed risk analysis tool in the modern world and it is called the Space Architecture Failure Evaluation (SAFE) tool. It is a Probabilistic Risk Assessment (PRA) tool, which addresses the physical risk of the space transportation system. SAFE performs Monte Carlo simulations of a system through its operational phases. The system is represented by its risk-driving components and a schedule of the state of the system. These components, along with a failure database developed as part of the tool, enable calculation of mean failure probabilities, uncertainty estimates, identification of the relative risk contribution of the systems and generation of risk intensity plots. The results allow designers to quickly identify high-impact areas for redesign or possible mitigation. Because the architecture is represented by its risk drivers, it is possible to perform high-level trades before all of the design details are finalized, impacting the design early enough to make changes if needed.[1]

For consumable products such as disposable goods, where the cost per piece is low, it is not that critical if, for example, the handle of a disposable cup comes loose. The cup can be easily replaced with a new one and the consumer may still choose the same trademark next time when buying disposable coffee cups. Instead, what may be critical for such products is whether it is appropriate for for mass manufacturing at speed. The model should therefore accommodate production-line machines so that no big investments are needed when there are changes.

Whatever the identified stumbling blocks are, the program management should concentrate fully on the critical path of the product program and make the necessary plans to tackle possible realizations of the risks.

7.2 Stumbling Blocks Specific to a Software Program

Every software product program tries its best to produce a high-quality product as fast as possible, i.e. at minimum cost. Is it possible to make quality at low cost? Is this not a contradiction? Not necessarily, but producing an error-free product within a relatively low budget certainly requires fully optimized working processes and very competent resources. Building such procedures and getting highly competent people on board of course needs investment, but once this investment has been made, the resources can also be utilized in future product programs.

Certain things and elements have proved to be common in almost all software product programs. This chapter introduces the most common stumbling blocks in today's software product programs, those that cause the program to fail to deliver what is expected within the given budget.

One software defect can have an enormous impact on a huge number of people. Understanding of the severity and importance of software failures has been globally recognized and is being intensely studied.

Construx Software Builders, Inc.[2] state that only 26 per cent of business system software projects are finished on time. How late the remaining 74 per cent of such projects are can be seen in Figure 7-1 The discussion in this section is also based on the same paper, which also introduces reasons why almost three-quarters of software projects either are late or cancelled.

Amrit Tiwana and Mark Keil[3] have introduced a quick and simple tool to calculate a project's collective risk value. The formula is shown in Table 7-1:

1. On a scale of 1 to 10, where 1 is low and 10 is high, how would you characterize this project compared to other projects in your organization?

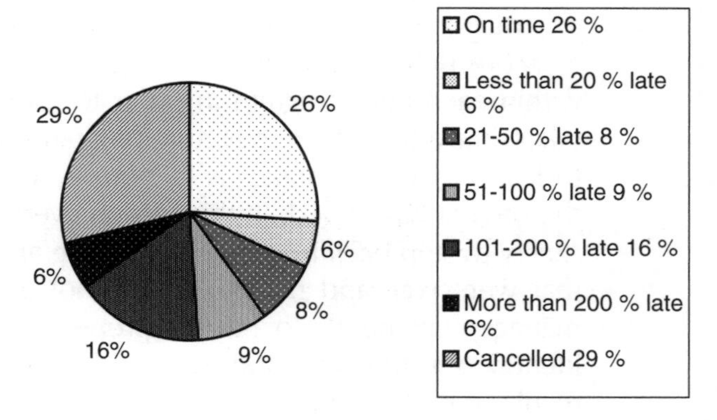

Figure 7-1. Typical project outcomes.

PROJECT CHARACTERISTIC QUESTION	1. RATING WEIGHT	=	2.
Fit between the chosen methodology and type of project	5 × 3.0	=	15
Level of customer involvement	6 × 1.9	=	11.4
Use of formal project management practices	1 × 1.7	=	1.7
Similarity of previous projects	3 × 1.5	=	4.5
Project simplicity /lack of complexity)	7 × 1.1	=	7.7
Stability of project requirements	9 × 0.8	=	7.3
3. Overall project risk score (higher score indicates lower project risk)			48

Table 7-1. The one-minute risk assessment tool.[3]

OVERALL RISK SCORE	PROJECT RISK LEVEL
10–28	High
29–46	Moderately high
47–64	Medium
65–82	Moderately low
83–100	Low

Table 7-2. Interpreting the risk score.

2. Add the six weighted ratings (see the worked example in Table 7.1).
3. A lower overall project risk score indicates a higher project risk with a range from 10 (most risky) to 100 (least risky).
4. Use Table 7.2 as a guide for interpreting of this score.

As this formula is very much simplified, it provides only trendsetting information on how risky or riskless a project is.

More detailed reasons for program failure should be identified within each new program in order to avoid making the same mistakes time after time in the future programs. One very good approach to learning from mistakes is to have a 'lessons learnt' session after the project has ended. In this session everybody should have a possibility to step back from the work done and analyse together things that went well and things that did not go too well. The program management needs to have a proper back-up plan for each risk identified in this way. Both risks and plans should be updated on a regular basis. A couple of already introduced ideas concerning software program risks and ways to control them are introduced in chapter 2. The following sections introduce the most common pitfalls in any software project.

7.2.1 Contradictory, Overwhelming or Too Many Requirements

Requirements can be excessive in many different ways. For example too many, and too demanding, requirements are naturally excessive. In addition, requirements that are technically impossible to implement are also of that kind. A technical feasibility study can save the program aiming too high and failing totally as the program turns out to be too expensive. Cancelling a program is not only financially frustrating, but it also demotivates the people in the program. In the worst case it may completely disable the whole organisation for some time.

Very rarely is the problem having too few goals in the project; too many goals are more likely. Sometimes the requirements can be technically or logically impossible to implement. Risk of that being discovered too late exists in every project. The less technical the people involved in the planning are, the bigger this risk is.

The goals in some cases are just too ambitious in comparison with the resources available and the expertise in the program. These things should be analysed by highly technical people, who have an inside knowledge of the development work.

The expectations of a product program, when compared with the available resources, both human and time, are very often too ambitious and unreasonable. Being able to identify this in a project can actually save the entire program.

7.2.2 Unstable, Incomplete and Informal Requirements

Another risk is ever-changing requirements. Sometimes this happens because the requirements were incompletely defined in the first place. Sometimes the customer wants to change the ultimate requirements in the middle of the program. Sometimes these changes are introduced within the program because, for example, market conditions change or a competitor releases a new version with a more attractive feature set. Whatever the reason, the following approaches to handling this kind of challenge are introduced in the Construx Software Builders paper *10 Keys to Successful Software Projects*:[1]

- user interface prototyping
- requirements workshop
- user interview
- use cases
- user manual as specification
- usability studies
- requirements reviews and inspections
- incremental delivery

7.2.3 Poor Planning and Project Management

Poor planning and project management have been identified as the second most common risks in software projects. A good project manager has good knowledge and experience in estimation of time and resources, life-cycle selection, quality assessment (QA) planning, technical staffing, project tracking, risk management and data collection.

One of the most important tasks of a program manager is to keep a constant eye on the state of the resources and on outputs. Evaluation of resource needs should be done regularly and should also be one of the very first and last activities in the program.

7.2.4 Unrealistic Estimates and Unjustified Expectations

On estimation, the Construx Software Builders say that the state of the art is dramatically better than the state of practice. Another practice worth trying to improve the quality and accuracy of estimates is to consider estimation as a mini-project. Periodically during the program the project management should re-estimate project characteristics. Unfortunately, this can sometimes be seen as a pointless task and a waste of money. However, it is the only way to guarantee readiness to act fast if the risk is realized.

7.2.5 Lack of Knowledge on New Technologies

Many projects tend to suffer from poor adoption of new technologies. A very new technology in a product automatically means a risk in a project. Implementing such a technology should be kept under special observation throughout the program.

7.2.6 Lack of Proper Risk Management

With active and competent risk management, it is easier to keep small problems from turning into big project-killing disasters. The more time-critical the project is, the more important is good risk management. In addition, if the program concerns an error-critical product, such as a medical device or space equipment, an absolute must is to follow rigorous and professional risk management practices.

In some areas, it is quite normal for a company to take risks, but if the company is beset by risks of all kinds, it may lose control over risk prioritizing. A key success factor for such a company is to separate non-strategic risks from strategic risks and keep control of the strategic ones. Risk analysis can also be viewed from a different point of view, as introduced in COCOMO/SCM Forum #17 Tutorial, 2002. Barry Boehm, USC.[4] A program should always decide 'How much is enough?' for the product and processes. What is the risk of doing too much versus what is the risk of doing too little? A program should tailor and adapt its life-cycle processes and determine what to do next. Boehm agrees that the risk management activity should start on day one. He also introduces a diagram showing the risk of delaying risk management in a program, which can be found as Figure 7-2.

The most critical risks are architectural ones as one unwanted 'feature' in the architectural specifications can have effects on system

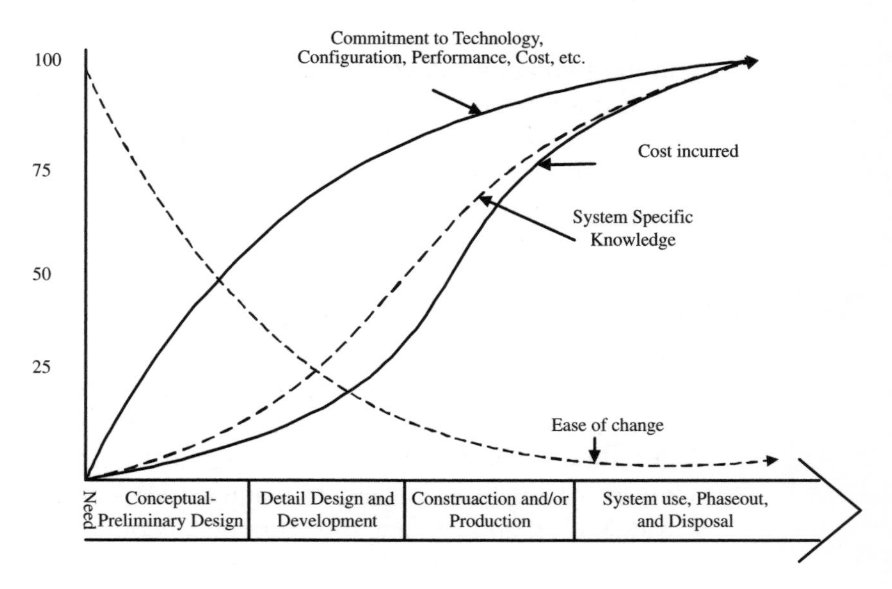

Figure 7-2. Risk of delaying risk management in a program.[4]

TECHNOLOGY XXX	
Risk description: A totally new technology is introduced for the first time in the program. People have no previous knowledge of it or experience working with it. The technology is important and therefore the success of the product pretty much relies on this technology being smoothly implemented into the product.	
Risk symptoms:	
1. Poor stability of the product 2. A large number of errors found 3. Implementation of corrections cause damage in other parts of the system 4. People get demotivated and frustrated if it does not work quickly	
Corrective actions:	
1. Strict gatekeeping of new fixes and implementation to be started 2. Prioritizing errors and fixing them one at a time 3. See above (point 2) 4. More resources to be involved or schedule to be replanned, people to be trained and no overtime permitted	
Probability: 0.6	Impact: Severe

Table 7-3. Example of a risk analysis table.

performance, interface integrity and other aspects such as adaptability and portability. Architectural risk analysis can be done in various ways, for example by reviews, simulation, modelling, prototyping and instrumentation.

One example of a risk analysis is to do it on paper, for example in the format introduced in Table 7.3.

It does not matter what sort of template is used in risk analysis as long as the results of the analysis are stored informally and reviewed on a regular basis. All of the elements introduced in Table 7.3 are good for inclusion in any risk analysis.

7.2.7 Lack of Organisational Integrity

As the average size of software has grown a lot over recent decades, more and more developers and teams need to be involved in a program before targets can be achieved. Organizational integrity is an absolute must for a successful business. Successful in this case means having the capability to manufacture products that are to be widely used and keeping the customers happy so that they want to choose the same manufacturer next time.

The things discussed in Chapter 6 on quality culture definitely have a key role when organisational integrity is considered. If the common goal is clearly communicated to all and widely accepted by those involved in a project, the targets are more probably going to be achieved.

Equally, the more people think that the company cares for them and is interested in their opinions, the better the results they achieve. Caring can mean, for example, providing people with the opportunity to attend training courses, treating them fairly and providing them with a pleasant working environment. All of these, of course, require some investment, but, on the other hand, if no attention as paid to this aspect, most likely the results will start deteriorating in one way or another.

7.3 Ways to Avoid Stumbling Blocks in a Software Program

Software professionals have invented several tools to avoid the most typical traps in a software project. These tools use different techniques such as algorithmic and parameter models, expert judgement, analogy and rules of thumb, to name a few. Unfortunately, each one of these suffers from a number of drawbacks. One tool relatively widely used for risk evaluation is the Constructive Cost Model (COCOMO) published by Dr Barry Boehm in 1981. Since then the nature of software project has changed quite significantly and therefore COCOMO needed to be re-evaluated. After several years of combined efforts by the University of Southern Carolina Center for Software Engineering (USC-CSE), IRUS at UC Irvine <http://www.ics.uci.edu/> and the COCOMO II Project Affiliate Organizations <http://sunset.usc.edu/research/COCOMOII/#researchsponsors> COCOMO II was introduced. COCOMO is widely used and several companies worldwide provide consultancy on it. For example, a company called SoftStar has published several papers on COCOMO. The following section summarizes SoftStar's overview of COCOMO.[5]

7.3.1 Overview of COCOMO

The COCOMO cost estimation model has been used by thousands of software project managers and is based on a study of hundreds of software projects. COCOMO includes the following:

- the underlying cost estimation equations
- every assumption made in the model (for example, the project will enjoy good management)
- every definition (for example, the precise definition of the Product Design phase of a project)
- the costs included in an estimate are explicitly stated (for example, project managers are included, secretaries are not)

COCOCO is seen to benefit it users in the following ways:

- COCOMO estimates are more objective and repeatable than estimates made by methods relying on proprietary models.
- COCOMO can be calibrated to reflect your software development environment, and to produce more accurate estimates.

The basic COCOMO model originally had a three-level hierarchy:

- **Model 1.** The basic COCOMO model computes software development effort (and cost) as a function of program size expressed in estimated lines of code (LOC).
- **Model 2.** The intermediate COCOMO model computes software development effort as a function of program size and set of cost drivers that include subjective assessments of the product, hardware, personnel and project attributes.
- **Model 3.** The Advanced COCOMO model incorporates all characteristics of the intermediate version with an assessment of the impact of the cost drivers on each step (analysis, design, etc.) of the software engineering process.[6]

7.4 Stumbling Blocks Specific to a S60-based Phone Program

A S60-based phone program is naturally subject to all those stumbling blocks introduced in earlier sections. In addition, it has some other special risks due to the industry and the nature of the platform. The following sections introduce some of the issues that need to be

Coding style and culture	Fixing speed	Testing activities and extent	Excessive requirements	Specification's sufficiency	Adaptation layer implementation
Baseline Selection					
Amount of Differentiation					
Testing Environment					
Integration Competence					

Application/ component level risks

Program level risks

Figure 7-3. Things to consider in a S60-based phone program risk analysis.

carefully considered in each product program, even if they do not seem to have be particularly significant. Figure 7-3 shows the elements of the S60-based phone program, which should play an important role in program risk analysis, in addition to those mentioned in previous sections. Each of the elements in the figure is explained in more detail in the following sections.

7.4.1 Program-level Risks

There are two levels of risk in a broad organization, where each team is solely responsible for implementing one or more specific components. First are the risks that should always be considered at the program level; if such a risk occurs in a program, it will destabilise the whole program and not only a limited number of components. This type of stumbling block should be recognized and analysed on a regular basis at the-program management level. The program level stumbling blocks are introduced in below.

7.4.1.1 Integration Competence

Overall system integration is seen as a program-level risk and therefore managing it in a proprietary way is essential in a successful product program. In a program in which the software development is that of a size appropriate to a smart phone, one of the three key activities is without any doubt the integration, and, to be more precise, when, how often, by whom and in which order it is decided the system should be built. Integration competence requires know-how on the order of the integration, i.e. which sub-systems are built first and what kind of impact one sub-system can have on other parts of the full software package functionality. Optimizing the integration requires knowledge of the internal dependencies of the components as well as practical-level knowledge on the usage of stubs and drivers.

Integration of S60 based phone software should happen stepwise and it should follow a certain order and structure, as described in chapter 12 on integration and the build environment.

7.4.1.2 Testing Environment

Testing as an activity is widely recognized to be one of the key activities in any product program. No testing can be executed without a proper environment. The nature of the product sets the requirements for the testing environment, but, whatever the product is, the testing environment is definitely a program-level risk.

The testing environment can turn out to be one of the issues causing delays in a S60-based phone program. This is especially the case if the environment required is considered for the first time after L1.2 (see Figure 3-2, showing the Licensee Milestones). Verifying a successfully integrated adaptation component task requires access to a live network and, if the location where the development takes place does not have network coverage, no tests can be run in a real environment.

The relatively massive requirements, both financially and technically, for the testing environment can really surprise the program, especially if all resources have been devoted to implementation tasks and no one has considered the environmental requirements in enough detail.

Plans for the testing environment should start at the same time as the test planning activities. For the majority of testing-related processes, test planning should be a parallel activity with the requirements management. This means that testing environment planning and budgeting should start more or less right after the very first product specification is available. Chapter 10 contains more a detailed description of the required testing environment.

7.4.1.3 Amount of Differentiation

A third potential program-level stumbling block in a S60-based phone program is an insufficient amount of differentiation. The more features a phone program decides to drop or disable from the platform, the bigger the risks are in getting the product stable. This is due to the internal dependencies of the components, as was explained in detail in chapter 2.

Each phone program naturally wants to have a terminal that contains something special compared to others. Specialities are produced either by dropping some features (in most cases some connectivity features) from the platform or by adding new pieces of functionality into the product. However the differentiation is done, it should be planned together with people who have a deep knowledge both about the internal dependencies of the platform and about how to exploit existing components and sub-systems in order to optimize the code.

7.4.1.4 Baseline Selection

Selecting the baseline wisely is one of the most important things when making a S60-based phone. Picking an arbitrary early platform version can create a need to integrate an extensive amount of fixes in-house. This makes it very difficult to integrate any fixes that come along with later releases of the platform.

As already mentioned in chapter 2, managing the program is by itself a relatively demanding task, as it can feel as though all the pieces in a multi-supplier environment are changing their shape all the time and the person in charge has to create a complete picture out of such pieces. Stabilizing the pieces as much as possible can dramatically ease the task. One way of stabilizing them is to avoid establishing one's own branch too early and making one's own fixes on the platform itself.

7.4.1.5 Defect Fixing Order

Very often a phone program, once it has discovered a set of defects in the code, starts fixing them without first prioritizing them. This can cause unwanted regression, especially once the code has reached its completed state. After the program has really reached a code-complete state, the program management should be very careful as to which fixes are to be integrated into the software. Tools and advice for doing the prioritizing of defects are introduced in more detail in chapter 11 on defect analysis.

7.4.2 Component-level Risks

Component-level risks are such that, if they occur in a program, they will not necessarily destabilize the whole program but instead cause additional problems in one or more components or sub-areas. Managing such stumbling blocks can take place in each team separately. The following sections introduce such component-level risks that are common to all S60-based phone programs.

7.4.2.1 Coding Style and Culture

MSW guides the customer programs in implementing uniform and easy to manage and maintain applications and solutions for specific markets. **Coding style** can be defined as *the way that the programmer brings clarity, maintainability, testability, reliability and efficiency to the code of a module*. This definition sets the objectives of good programming style, but what it does not do is to define whether a piece of software is good or bad style. No matter how perfect the software design is, the final product will be expensive to maintain and test if its implementation, the code, is of poor quality.

Some of the following principles are very general and therefore applicable to all software projects. Equally, they should be considered in a S60 based phone prior in order to get a fine product out as planned.

- **Reusability.** Avoid rewriting and copying verbatim code written by someone else. If you feel the need for a common module, communicate this to the whole team. In this way, the code size and ROM consumption can be reduced, which leads to cost savings in hardware and production time.
- **Maintainability.** One of the good qualities of good code is that it is understandable and easy to read. Sometimes the original programmer has moved to a new job and the maintainer has no history with the implementation.
- **Modularity, Encapsulation and Information Hiding.** If the code of a module becomes very long and complex, whether the functionality should be re-organized should be checked. Constant monitoring of the module sizes could be worth doing, especially in the most critical areas.
- **Assumptions about the user of the code.** The implementation should be done so that it can protect itself from possible misuses. As a minimum, documenting all restrictions is necessary if they cannot all be implemented.

- **Commenting the Code.** Simple comments can easy the maintainability of the code. However, if the code needs to be explained in detail, then it is probably not clear enough.
- **Modifying the code.** If the maintainer of the code is different from the implementer, the maintainer should use the same coding style or, if that is not possible, then change the entire coding style to new one.
- **Compilation.** There must be a well-known and clearly justified reason to ignore warnings during compilation.
- **Internationalization issues.** It is very important to write code that can be easily localised to different languages without having to make major engineering changes. The main principles in keeping the code easy to localise are:
  - Keep code and content separate
  - Use Locales
  - Allow for test expansion
  - Do not concatenate
  - Do not reuse strings
  - Use re-orderable parameters in strings
  - Do not use test in graphics
  - Comment the text strings
  - Follow the formats and use the templates provided

  In addition to above, the licensee will receive an internationalization guide along with the deliveries; it is highly recommended that programmers study the document carefully.
- **Symbian-specific things**. The Symbian OS is designed to contain a highly functional application in a resource-constrained environment. Robustness is the key for the end-users' acceptance. The Symbian OS resource management and cleanup framework provides the needed robustness and scalability that is unparallel in the application sector. Therefore, every programmer should be familiar with the cleanup and memory management in Symbian OS.

- **Miscellaneous.** The following list of coding hints represents some guidelines for development in the Symbian environment:
  - Do not use any white spaces or non-ASCII characters in filenames.
  - Do not rely on case-sensitiveness in filenames.
  - Avoid references to physical file system paths.
  - All new classes should be stored in separate files.

  In cases where multiple clients need access to a shared resource, a client – server interface needs to be implemented.

  - Isolate machine- or compiler-dependent code sections into separate files.

7.4.3 Fixing Speed

One of the most significant stumbling blocks in S60-based phone programs has been the surprisingly large amount of regression after the code-complete stage. No single root cause for the regression has been identified but it seems that a combination of several unfortunate false actions simply decrease the overall manageability. One can cut corners and say that regression is generated by uncontrolled integration of new fixes into the builds. In such cases regression could be avoided by taken a strict *gate-keeping* process into use. The program maturity curve in Figure 7-4 shows how the system maturity evolves in comparison with time and what sort of activities should be planned and when.

In cases where the gate keeping is neglected partially or completely, the curve after code-complete can become jagged. In the worst case the maturity of the system can collapse to a level that correlates with the maturity at a very early phase of the implementation. In such a case some features can be blocked and reverse engineering the situation to find out which fix or fixes caused the sudden loss of stability can be technically very demanding and time-consuming.

7.4.3.1 Testing Activities and Extent

It can be surprisingly difficult to determine how much testing is enough. The average stumbling blocks around testing activities can be divided into three:

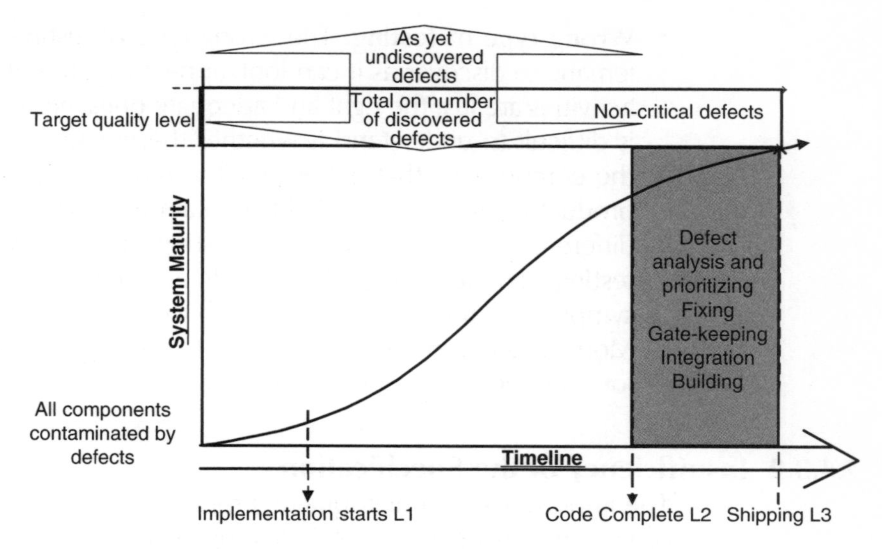

Figure 7-4. System maturity evolution.

- **Too extensive testing.** Too extensive testing means that, because of the lack of technical judgement on which parts of the code are either the most important or the most faulty can lead to a situation where too many resources are tied to different testing activities. One symptom of this is the relatively low test-case hit-rate. The way to manage this is to check the percentage hit rate of tests, i.e. how many of the test cases run actually found a defect. If the percentage is very low, it may be worth analysing the current testing activities and trying to find a more efficiency test procedure to follow.

- **Too little testing.** Too little testing means that for some reason (for example, that the program may have too optimistic assumptions about the quality of the product) the planned testing activities are not able to discover the defects that need to be found. The risk with this is that if an unstable product is shipped, in the worst case it can harm the reputation of the entire company. It is very difficult to say if planned testing activities are too light or not; one way of finding out is to set clear targets for testing, such as the number of defects it needs to discover, and then check whether the target is achieved or not.

- **Wrong type of testing.** The wrong type of testing is very problematic to discover as it can look appear that the planned testing activities are just the right and adequate ones. Nevertheless, what is difficult to understand is whether these activities are checking the components that either play the most significant role in the product or otherwise probably contain many defects. Seeing the difference between too much testing and the wrong type of testing can be somewhat difficult as they can have a similar symptom, a relatively low defect hit rate in the executed tests. More detailed information and hints on proper testing activities for a project is provided in chapter 9.

7.4.3.2 Insufficiency of the Specification

The importance of a high-quality product requirement specification of course applies to all product programs and is not specific to a S60-based phone.

The most critical component and sub-system interfaces need to be documented in a very detailed way to ease interoperability and integration. One way of checking whether this problem is occurring is to make an early integration round just to see whether all teams are going in a common direction. This, of course, does not discover any incompatibilities implemented in a later phase.

7.4.3.3 Adaptation Layer Implementation

In order to make the S60 platform software run with the licensee's cellular hardware and software, special adaptation software is required. The adaptation software integrates the S60 platform on the underlying cellular platform. This software is referred to as the adaptation layer. The successful implementation of this rather big layer can sometimes be a stumbling block in a project. Those who are to be given this task need to know both the platform telephony API and the licensee-specific modem software.

S60 licensees need to implement the adaptation layer, although some reference implementations are provided along with both Symbian OS and the S60 platform.

The adaptation layer consists of the following parts:

- Provider Modules
- Hardware-specific Symbian software

7.5 Provider Components

Provider modules are licensee specific software that link the S60 platform services to customer terminal cellular software such as Domestic OS software.

7.6 Summary

Each product development program has its specific challenges. Very often these challenges come from the organization, but the product itself can also bring some special issues to the program. S60 brings special potential stumbling blocks to the program that will cause problems if they are ignored. The S60 architecture is relatively complex and this, combined with the fact that customers get the earliest versions of the platform, means that the challenges are focused on how to manage the integration. Some tools for extensive program-level risk analysis should be used to at least identify potential risks and prepare for the risk to occur.

Chapter 8: Platform Testing versus Platform-based Phone Testing

Some decades ago testing was often considered as a useless, time- and money-consuming activity in a software project. However, since the sizes and complexities of the average software program have grown along with the time, testing has become recognized as one of the three main activities (together with design and implementation) in all software projects throughout the world. This turn-around point occurred approximately three decades ago. Nowadays, no self-respecting organization will carry out any significant implementation until the required test plans have been documented, reviewed and approved.

As shown in Figure 8-1, S60 full delivery includes a full set of documented test cases. These test cases describe the tests that have been run on the platform, or are to be run on the platform prior to the final delivery of the release. In part, these documents are delivered in order to make the testing quality visible to the customer so

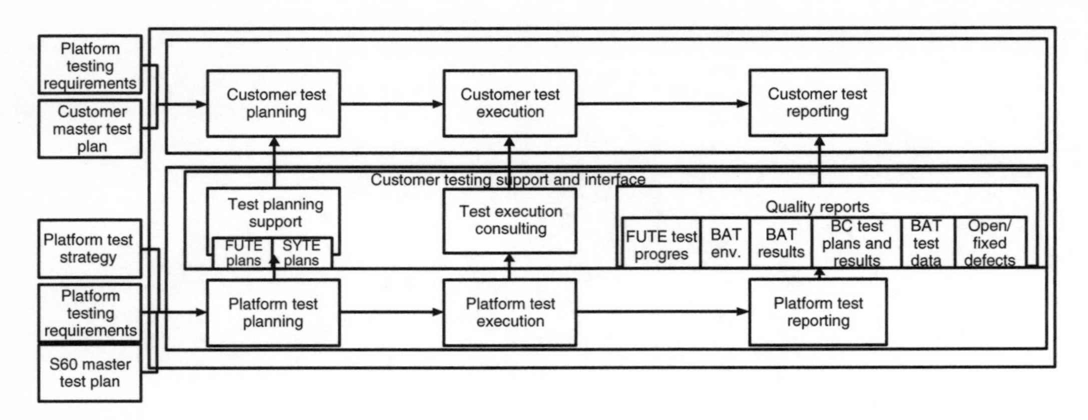

Figure 8-1. Customer testing support.

that the customer can decide how probable it is that their program inherits new defects from the platform. This chapter focuses on platform test planning and the execution processes.

The platform-based phone program will find that some of the tests that come along with the platform delivery are worth copying to the phone program test plans. However, the very important difference between platform testing and platform-based phone testing should be fully recognized.

8.1 The S60 Testing Process

As described in chapter 2, the platform development follows both incremental and iterative methods. This also applies to testing. Once a new release program (for example 3.1) is started, the feature set is frozen and the tests necessary to guarantee visibility on the quality are planned. Test planning can take place in parallel with implementation up to some point and it is normally completed by E2, after which the customer programs will have received all test documentation. Figure 8-2 shows the overall Test Management process in S60.

In addition to the documentation and consultancy delivered to the customer, some test classes are also included in releases. These test classes can be utilized in the customer program either by ebing exceuted as they are or as a basis for the customer's own test class design and implementation.

8.1.1 Platform Test Execution Process

Platform testing follows the most commonly known testing phases on a one-to-one basis. These phases follow the ISEB 7925-2[1] stan-

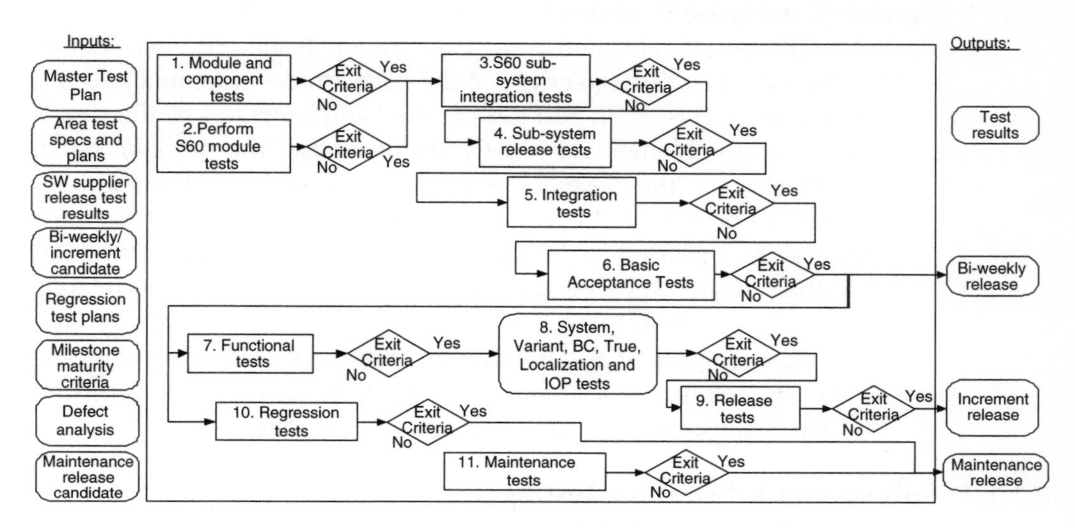

Figure 8-2. *The S60 test management process.*

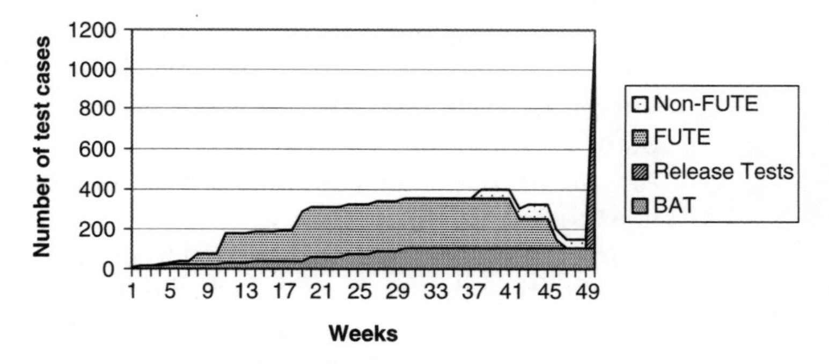

Figure 8-3. *S60 test execution example.*

dard activities, even though some of them might be given a different name. However, since the ISEB standard is the only testing terminology standard in the world, it is used as a reference point in this chapter when describing the platform testing activities.

The progress of the platform testing procedure follows the pattern shown in Figure 8-3. The Basic Acceptance Testing (BAT) is the only testing done on every individual bi-weekly release. The rest of the couple of thousand of cases are run only once during the program if the following two things are true: first, the test is passed and, second, there is no further risk of regression caused by fixes to any other components that would make it necessary to re-run this test case.

8.1.1.1 Module/Component Testing

Module testing, also known as Component Testing or Unit Testing, is the very first testing activity in the platform development and it is done for all classes. In S60 it is planned and executed by the developer, who implements the code. In most cases the module testing takes place in the developer's PC by utilizing stubs and/or drivers that have been coded in separate classes. In some cases these 'programs' are included in the platform deliveries. The only instruction on their usage can very often only be found in the comments relating to implementation.

Each component and module needs to pass a preliminary testing criterion before it is accepted for integration. This criterion is related to two issues: testing code coverage and test pass rate.

8.1.1.2 Sub-system Integration Testing

After the component or module has passed the module-testing phase, it is pre-integrated with other components to make a bigger entity called a sub-system. This sub-system is tested within the integration testing activity. Since the entity only represents one part of the whole software package, it very likely needs either stubs or drivers, or both, around it to act as neighbour components. The purpose of integration testing is to check how well the component's APIs are implemented and how well two or more components work together.

Integration testing is only applied to chosen sub-systems that are found to be either most critical or otherwise risky to the entire software. Those that go through the sub-system integration testing phase also need to fulfil certain predefined criteria.

8.1.1.3 Basic Acceptance Testing (BAT)

Basic Acceptance Testing (BAT) represents the alpha acceptance testing in the ISEB-standard. It is a relatively small sub-set of all functional test cases and it is the only test set that is executed on every software packet before delivery. BAT covers all features but is a very light test package, as the ultimate purpose of it is to provide first-hand information on how successful the build process was and to discover any possible blocked features in the build. In the other words, one can consider BAT as a kind of sanity check tool that provides quick and dirty information on the success of the build. To save time and avoid unnecessary testing activities, the number of BAT test cases should be kept relatively low.

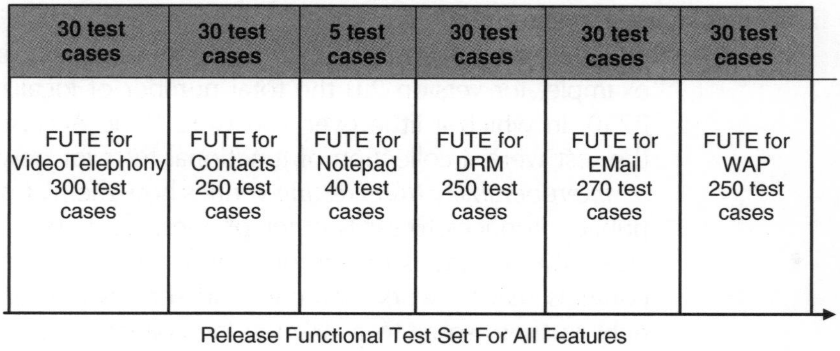

30 test cases	30 test cases	5 test cases	30 test cases	30 test cases	30 test cases
FUTE for VideoTelephony 300 test cases	FUTE for Contacts 250 test cases	FUTE for Notepad 40 test cases	FUTE for DRM 250 test cases	FUTE for EMail 270 test cases	FUTE for WAP 250 test cases

Release Functional Test Set For All Features

Figure 8-4. BAT test set.

BAT cases are delivered to customers together with the results of the BAT test round.

Figure 8-4 gives an example of the extent of BAT cases in a platform release. The numbers are not exact.

8.1.1.4 Functional Testing

Functional testing (FUTE) is the same as its namesake in the ISEB standard. The purpose of FUTE is to discover errors in code that cause a malfunctioning of some platform feature or application. S60 FUTE cases are planned by utilizing predefined use cases and use case scenarios. The majority of these tests are communicated to customers. For all platform releases the number of FUTE cases has grown to be over ten thousand test.

Functional test cases are run sequentially. Since, for example, the Contacts application is most probably ready before, for example, telephony, the functional tests of the Contacts application are run before those for telephony and for a different release or build.

8.1.1.5 System, Localization, Binary Compatibility and Interoperability Testing

System, localization and binary compatibility represent non-functional tests in the ISEB world, where interoperability generally relates to integration testing.

System testing concentrates on overall platform performance, power management, memory, stress, volume and speed. *Localization testing* focuses on finding possible problems in localization builds, because localizing a build to different languages very often causes unpredictable changes. An average 35 per cent of the totally new features of a platform release's functional tests and BAT are therefore included into this testing activity for each localized build. As an example, for version 2.0 the total number of localization tests was 2250, in which a little over 100 were Basic Acceptance Tests and the rest were a collection of functional tests for new features.

Interoperability (IOP) testing verifies how the platform implements public interfaces towards other phones, networks, servers and services. IOP testing is mainly done in a laboratory environment that contains real network elements and servers plus an administrative right to them so as to guarantee log-collection for further problem solving. There is more discussion on the IOP test environment in chapter 10.

8.1.1.6 Release Testing

Release testing is a sub-set of functional tests, localization and system tests. For version 3.0 it includes over 1000 tests and its purpose is to verify whether a release is ready to be called an increment or not. It provides knowledge on whether or not the release has the maturity an increment needs.

8.1.1.7 Regression Testing

Regression testing takes place if some fix potentially carriers a risk of damaging some other functionality. Regression testing is a collection of different carefully selected test cases. It consists of BAT and fixes specific functional test cases plus, of course, a fix-specific verification test.

8.1.1.8 Maintenance Testing

Maintenance testing is done once the platform is in maintenance mode, i.e. once it has been proved to be at a commercial quality level and is publicly available. Maintenance testing consists of two parts: BAT and fix-specific extensions, which can be regression test cases or just fix-specific test casse.

8.1.1.9 S60-based Phone Testing

Once a customer gets a platform good enough to be used as their baseline, they also get the information on what kind of testing the release has undergone before delivery. It is quite natural that exactly the same test set is not worth carrying out because most results will remain the same as during testing of the plain platform. Customers should therefore consider carefully the extent of testing in their pro-

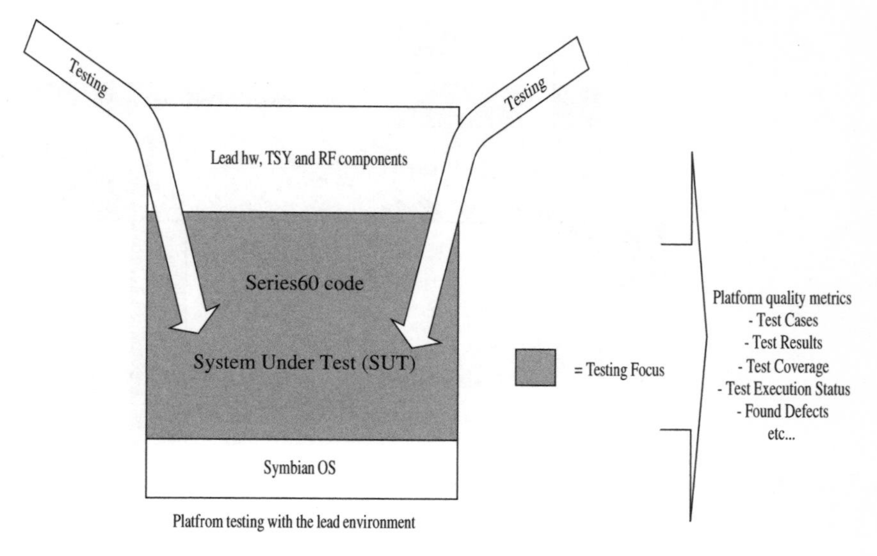

Figure 8-5. Platform testing targets.

grams and avoid extra testing that does not bring any new information to the program. Figures 8-5 and 8-6 show the difference between platform testing and platform-based phone testing.

Figure 8-5 shows that the platform is thoroughly tested with the help of the lead environment. The lead environment is a chosen phone program that is developed in parallel with and with the help of the platform program. Any customer program can become a lead program for the platform release if the program fulfils certain criteria. The criteria include requirements on things like the number of prototypes availability to MSW. The maturity of the lead terminal increases along with the increase in the platform maturity. The lead environment provides the otherwise missing parts of the full product, such as telephony components, the radio-layer, hardware, etc.

Figure 8-6 shows how testing should be targeted in a platform-based product program. Clearly there is very little risk in the platform components already proved to work by MSW and having very few dependencies on other modified or otherwise risky phone components.

To simplify a little, every program should do their own test planning, which can be optimized by taking the required test cases from platform deliveries and running them together with the program-

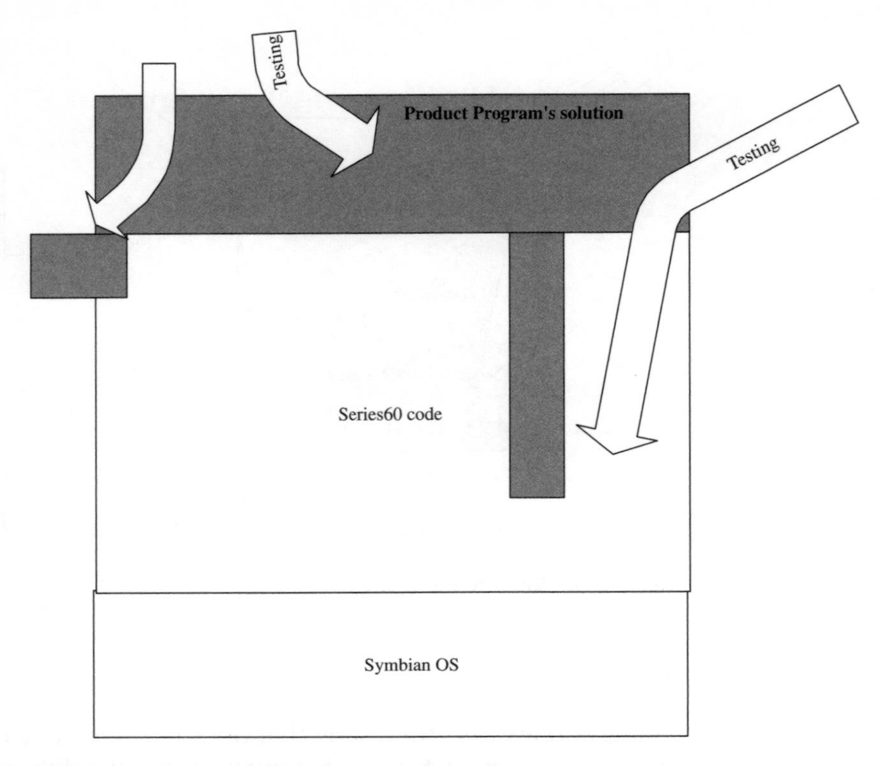

Figure 8-6. Platform-based phone-testing target.

specific test cases. Optimizing a healthy balance between these two can be tricky, but, if it is done properly, guarantees that no time is wasted on useless testing and the same time provides a full understanding and visibility of the maturity and defects of the phone.

How can an optimum usage of a given test cases be approached? The following three types of planning need to be in place and followed for the success:

1. Planning based on architectural analysis
2. Planning based on baseline maturity analysis
3. Planning based on fix analysis

Planning Based on Architectural Analysis

This includes areas in system architecture that

- have most dependencies on other sub-systems
- have been modified by the customer program
- play a critical part in a system

These areas also need to get most attention from a testing perspective too. On the platform side the testing has already been prioritized so that the most critical parts get to be tested most, but if a customer program affects these implementations, they also need to be given additional attention; either enough test cases should be picked from the platform delivery for re-running or the customer program should plan their own test cases. More information on this topic is included in chapter 9.

8.1.1.10 Planning based on Baseline Maturity Analysis

As already explained in chapter 2, choosing a baseline should be a result of analytical thinking and technical intelligence. The following two things should be taken into account:

- release quality, overall run rate so far and the number of defects found; also the minimum number of fixes that need to be integrated into the previous release of the system
- timing, i.e. how much longer can a product program wait until it has a solid platform on top of which to build the product

As shown in Figure 8-7, the later the baseline is selected, the smaller the number of fixes that need to be integrated into the system. Yet, at the same time, this decision must take into account the cost of guaranteeing the correctness of the APIs.

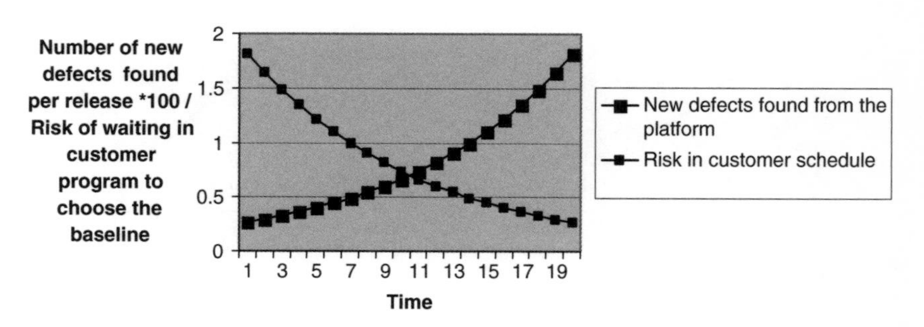

Figure 8-7. Baseline impact on test planning.

8.1.1.11 Planning Based on Fix Analysis

Fix analysis plays probably the most important role in successful test planning. More detailed hints and tools for a proper fix and defect analyses are to be introduced in later chapters, but, because it plays an important role in test planning, it is also introduced here.

No matter where the fix originally comes from (Nokia, subcontractor or in-house product program), there are two things at stake:

- the risk that this fix will cause either too many new general defects or unwanted regression concerning the critical features
- the risk that omitting this fix from the system will cause to customer satisfaction, to the company's reputation and the company brand

8.2 Summary

The platform goes through an extensive testing during the development program. Testing is done with the help of a lead product, which provides the hardware and modem software to the platform program. Once the platform reaches commercial quality, it has passed thousands of test cases and, in that way, proved to fulfil and even exceed predicted customer expectations. For the sake of time and money, the S60 customer program should not run all the same tests again; instead it needs to focus testing on the areas that have been modified or added or otherwise feel uncertain in the product. Test planning techniques based on the architecture, baseline maturity and fix analysis are all recommended processes to be followed in any S60-based customer program.

Chapter 9: Testing as a Tool

It has been said that there is not a single line of code with no defects and the many implications of the word quality have already been considered in earlier chapters. The ways to make high-quality products were also explained earlier. Even though these suggestions are worth following, they do not provide enough information. Software programs have a special nature; no matter how defect-free is the code one tries to implement and no matter how well quality-oriented the organization is, everyone who programs creates defects while coding. This chapter and the following chapters provide information about tools that help to create software that is more error free in general. However, it is important to understand that testing by itself can never improve product quality. There are unfortunate examples where the testing team has been blamed because they found too many defects, as if they had had something to do with the fault creation. The testers' role is very simple: they create awareness about product maturity, nothing else. Another interesting fact is that defects seem to address themselves to the same components. Roughly speaking, 20 per cent of the entire system can contain 80 per cent of all defects.

S60 Smartphone Quality Assurance Saila Laitinen

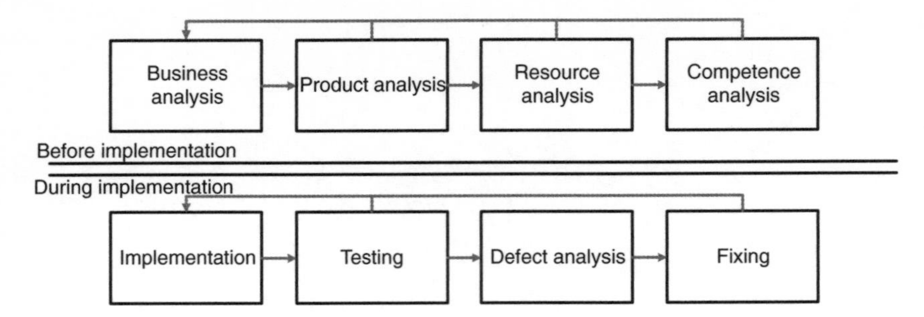

Figure 9-1. The phases in Quality Assurance.

Figure 9-1 shows an overall plan of the activities that are needed in high-quality product implementation. The dark grey boxes in the figure are those that are discussed in the upcoming chapters.

It is very difficult to decide which way to go unless you have a clear vision on where you are standing now. In other words, good testing reveals the current weaknesses and strengths in a product; it makes the quality visible at a certain point in time. By analysing the test results, the project management should have a better view on what is needed to make the product complete. All of this applies to software programs too. There are three things needed to create software that is more error-free:

- Implement the product so that the number of created defects is minimized.
- Make quality visible by testing, i.e. discovering in a timely manner the most important defects generated in the product. This requires testing covering most effectively those features that are vital for end-users.
- Improving quality by fixing in a controlled manner, i.e. so that the regression is minimized.

The concept of testing is a subject that is discussed throughout this book. At the beginning of an informal design process, the approved requirements can be very easily misunderstood, which can produce defects in the final product. Product quality improves if the contradictions in a product description, or between two such descriptions, are successfully eliminated. In order to carry out very high quality testing, the program has to understand the root causes of defects in the product.

The structure of this chapter is as follows. The first part introduces testing in different development processes. In the second part different testing techniques and tools are introduced, while in the last part the different testing phases are explained one at the time.

9.1 Testing in Different Processes

All testing activities need to fully take into account the development process. There are certain rules that apply to every product development program such as a software program. The V-model in Figure 9-2 was developed to regulate the software development process within the German federal administration. It describes the activities that take place and the results that have to be produced during software development.

The challenge comes when the requirements of an informal design process are misunderstood in later phases of the program. That is why all intermediate products such as documentation should be tested against the outcome of an earlier stage. Figure 9-3 introduces the activities that can prevent these defects from remaining undiscovered in the product. But how can tester test documentation? As boring as they sound, formal reviews are a good approach to discovering defects. In a good review, a group of people read the documentation and try to discover all the illogicalities in a single

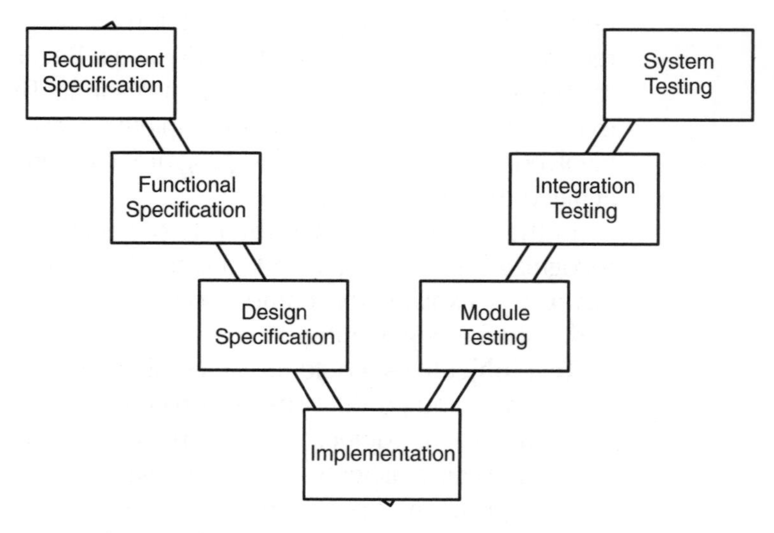

Figure 9-2. The V-model.

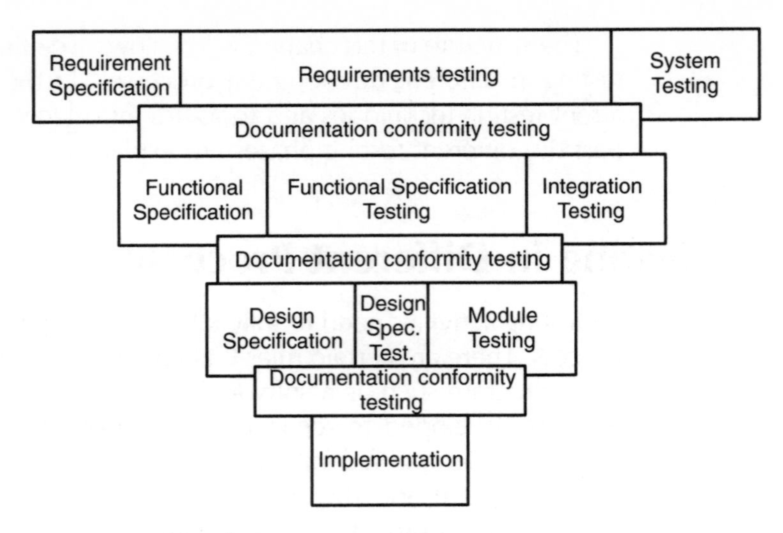

Figure 9-3. V-model with the testing activities.

program specification and the contradictions between two or more specifications.

The V-model needs some updates to become a more complete model. These updates are documentation conformity testing between each pair of activities and specification tests carried out in review sessions.

Sometimes the program management carries out the documentation conformity testing but in an informal way and they may not recognize it as being a testing activity at all. Because of the informality, this activity tends to have some holes in it. There are some tools that can help in achieving more complete reviews. Even if the people involved are competent, the program can probably still help to achieve the best results.

In the mid-1990s, when the role of testing became more widely recognized, Rick Craig introduced the term 'testware engineering'.[1] Figure 9-4 explains the comparison between software engineering and testware engineering.

Test objectives could be a combination of a company level *Test Strategy* and product-specific objectives, which specify the target in terms of the defect hit-rate, resources, processes and other product-specific aspects. Processes play an important role in any product program. If processes are too formal, they can decrease the innovation and overall flexibility in a program, while, on the other hand, if they are too informal they can lead to a situation where

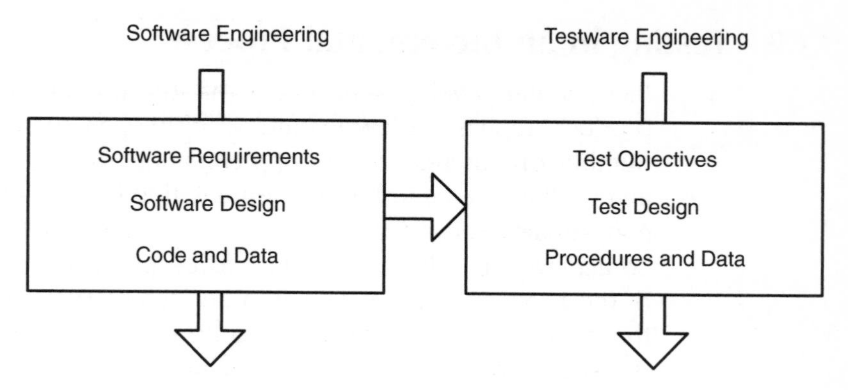

Figure 9-4. Software Engineering versus Testware Engineering.1

the original requirements are misunderstood during some phase of development.

Well accomplished testware engineering can be very demanding. The things that make it demanding are the following:

- the complexity of a software product
- the nature of a software product
- the nature of the problems in a software product
- the amount of information and know-how needed

In addition, people-related issues, such as frustration with the documentation, schedules, requirements, changes and attitudes, make it very challenging.

Chapter 6 explained different development processes, while this chapter explains in more detail testing in each of these processes.

9.1.1 Testing in an Iterative Process

In the purest iterative process, at least in theory the same resources can be used for both coding and some testing, because these two activities are more or less sequential. The techniques used in iterative development depend fully on the testing phase. Testing in an iterative process is quite well organized and easy to manage. The bigger the product program, the more testing teams there are involved. A program can have separate test teams for module, integration and system testing activities. Each of these teams can act rather freely if the exit criteria for each test phase are well defined.

9.1.2 Testing in an Incremental Process

Incremental development means that the program implements the product stepwise, a few features at a time. Maturation of a limited number of features prior to implementing any other features guarantees that, even if further implementation fails to deliver, the earlier good-quality set of features can be used and put onto the market. To achieve this, all features need to be prioritized at the beginning of the program. The challenge in an incremental process comes in the parallel activities that need to be managed all the time. Figure 9-5 shows these activities in a timeline.

9.1.3 Testing in an Agile Process

Most agile methods attempt to minimize risk by developing software in short timeboxes, called iterations, which typically last one to four weeks.[2]

Since agility emphasizes real-time communication, the testing personnel should sit near the developers and the overall process should focus on effective face-to-face communication over formal documentation. The agile process is the most flexible process when it comes to the order and punctuality of activities. It sets extra requirements for project follow-up and management while providing potentially very good results. This process sets specific requirements on testing. Testing must always be prepared well in advance, but be ready to be modified and to start whenever the need occurs.

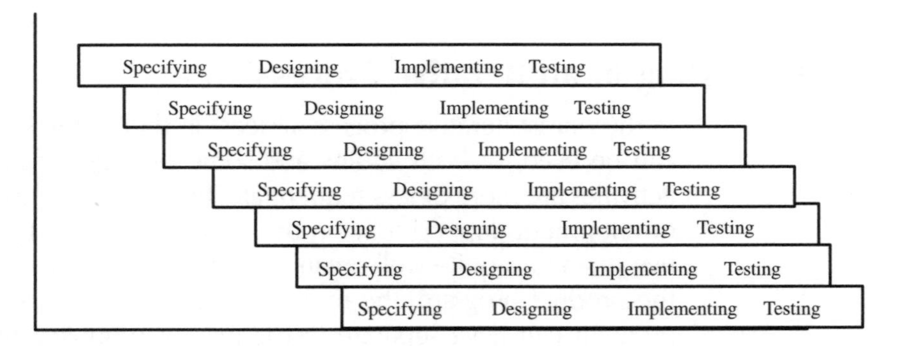

Figure 9-5. Testing in an incremental process.

9.1.4 Testing in an Extreme Programming Process

Extreme programming (XP) was first introduced in the early 1980s. Testing has quite a singular role in XP. XP does things in reverse; the tests are written first and after that the code is implemented. The coding is finished after it has passed all tests and the programmers cannot think of any more tests. XP aligns well with adaptability but the trade-off is predictability. The benefit of XP is that the programmer really must think and understand what he or she is about to implement before carrying out the coding.

9.2 Testing Techniques

Since the software developer or software architect creates defects unintentionally in an unsystematic way, the testing should be an intentionally systematic activity. Early test preparation should detect errors from the corresponding construction phase. Defects could and should be detected when tests are prepared, not when they are run. In this way the program can decrease the risk of not being able to keep to schedule or budget, because the earlier a defect is fixed, the smaller the potential risks. With this approach the program can potentially save significant amounts of money and time and still come up with a final high-quality product. Very late testing phases simply cannot find all the defects in the product. Unfortunately, this does not mean that end-users will not find them. It is not very unusual for the project to decrease the time planned for testing as a result of the development phase being prolonged. Too many times we have seen where this kind of approach ends up.

Testing efficiency means that the tester detects the most significant defects first with a minimum test effort.

The success of testing is not only dependent on the testers' skills but also on the testability of the product. Testing a bad product is a waste of time, because not enough defects are found anyway and there will never be enough time to fix all those that are found. Some products are more testable that others. Testability describes how easy it is to test a product. It is a composition of two things:[3]

- **controllability:** the ease of setting all data used by the program into a known state
- **observability:** the ease with which all relevant details of the program under execution can be observed

Things that decrease a product's testability are:

- average function/class length
- cyclomatic complexity
- dependencies
- usage of unnecessary data

There is only one method of finding nearly all defects: long-term use by many users. However, the goal is not to find all the defects, but to find as many errors as possible and, in particular, those errors that it is economically wise to fix.

There are numerous software testing techniques that have been introduced over the years. Some of these techniques have undeniably become obsolete as software processes have evolved and changed. In addition, the terminology around software testing is very wild and colourful. This section introduces some elements of basic testing techniques. The vocabulary used is based on the ISEB Standard number 7925-1.[4] Testing can be constructive or destructive depending on the tester's mindset and targets.

Constructive testing means that testing is trying to prove that some piece of functionality works and not to show that it does not work. Constructive testing is very often based on predefined use cases. In the most blatant case, constructive testing is effectively proving to the potential customer that the software works. Constructive testing can be a very risky choice if it represents the only testing approach in an entire program.

Breaking the functionality by doing things to the product that can be described as hacking is called **destructive testing**. Destructive testing can either be well-planned or be a very ad-hoc type of activity. The benefit of so-called 'monkey testing' is a relatively high hit rate in discovering defects. The drawback of unplanned testing is the difficulty in documenting the actions that caused the failure. This can rebound during fixing, where understanding the root cause is vital.

White-box testing (also known as glass-box, structural, clear-box and open-box testing) is a software testing technique whereby explicit knowledge of the internal workings of the item under test are used to select the test cases and data. Unlike black-box testing, white-box testing uses knowledge of specific programming code to examine outputs. The test is accurate only if the tester knows what the program is supposed to do and how it does it. He or she can then see if the program diverges from its intended goal. White-box testing does not account for errors caused by omission, and all visible code must be readable. It means that the tester has a clear knowledge of how the software is built and how it has been implemented. The benefits of white-box testing are the predictability concerning potential defects, although sometimes this can also turn out to be a restricting factor. The earlier the testing phases in question take place, the more they utilize white-box techniques.

Black-box testing (also known as functional testing) is a software testing technique whereby the internal workings of the item being tested are not known by the tester. For example, in a software-design black-box test the tester only knows what the inputs are and what the expected outcomes should be and not how the program arrives at those outputs. The tester does not ever examine the programming code and does not need any program knowledge other than the specifications. This means that the tester does not need to understand the software design or the implementation. The only thing that is known is the requirement on how a certain feature should function from the user perspective.

Defect Amount Estimation is a recommended technique for use in focusing tests on those components and sub-systems that most probably contain the most defects and the most critical defects. Such sub-systems are, for example, those that implement one or several of the following 'rules':[3]

1. It is easy for programmers to become confused as to just what a **pointer** is pointing at. Manipulation of pointers often requires an understanding of, and dependence on, the underlying processor architecture.
2. Dynamic memory allocation and de-allocation are often closely connected with the use of pointers.
3. Unstructured programming, including the use of GOTO is perhaps the most widely recognized source of programming failure.
4. Multiple entry points and exit points, loops, blocks and functions are really just a variation of unstructured programming. However, there are cases in which carefully controlled use of more than one exit can simplify code.

5. Variant data, meaning cases where the data in a variable changes or the structure of a record changes is difficult to analyse. It can easily be misunderstood and lead to a programming error.
6. Implicit initialization: a simple spelling mistake can result in software that compiles but does not execute correctly.
7. Concurrency and interrupts can be source code problems. It is easy to forget about parallel execution when designing or coding a limited number of components.

Defect seeding can be used to verify the extent of testing. In defect seeding the programmer purposely adds defects into the code before testing. This is based on a concept where the seeded defects are populated homogeneously throughout the product in the same way as the other non-intentional defects. The ratio of the number of seeded defects detected to the total number of defects seeded provides a rough indication of the total number of unseeded defects that remain undetected:

IndigenousDefectsTotal = (SeededDefectsPlanted/ SeededDefectsDetected) × IndigenousDefectsFound

Although defect seeding can only provide an estimate of the remaining unknown defects in the product, it is mostly used for academic research purposes.

Estimated Degree of Functional Usage (EDFU) is a numeric value that provides a simple estimate of how often the end-user uses a certain piece of functionality. To determine this one needs to understand consumers' behaviour very well. If the value is one, it is estimated that the end user uses this particular piece of functionality every time he or she uses the product. If the value is close to zero (for example,. 0.01), it is very unlikely that an average user ever uses such functionality. A good question to ask at this point is how to define an average user. In the mobile phone industry devices are targeted to certain customer groups. This helps end-user understanding a little. In any case, the program needs to create a customer profile based on the price, feature set and when the product is available on the market. Then the program needs to find a group of people that match this profile and either give these people the first version of the prototype product for use and watch how this potential customer group uses the product or, if a prototype is not available, ask the group to complete a questionnaire on how they think they would use the product.

Naturally, the higher the EDFU, the more important it is for that part or function to work properly. However, even such features that have a value 1 for EDFU are not automatically equally important from a testing viewpoint. Those features whose implementation results in architecturally complex solutions should receive the most testing attention.

Data-driven testing aims to find defects in which certain data is incorrectly processed. It focuses on every data area of interest in a product. Equivalence class partitioning, boundary value analysis, domain test, special value test, category partitioning test, dependency test, random test and syntax test are all types of data-driven testing. For example, using values above a higher boundary or below a lower boundary can be very effective in finding defects. In addition, using correct or wrong data types and special values can be a good testing approach.

Logic-driven techniques try to identify all incorrect handling of the logic. This can be done by, for example, testing certain combinations of inputs to every logical expression. Logic-driven testing examples are, for example, testing every condition using cause – effect graphing and doing meaningful impact strategy or minimal multi-condition tests.

Event-driven testing discovers incorrect handling of events. The time sequence and arrival time of different inputs may introduce failures. For two input events, testing only with event 1 or with event 2 may be worth doing. In addition, the time distance between different arrivals may be changed. The following events are good when testing time-outs:

- arrival before timer is set
- arrival before time out
- arrival at time out
- arrival after time out

State-driven tests aim to find all incorrect state transitions. They are always based on state transition diagrams. All critical states and critical transitions need to be tested. Trying to execute combinations of transitions is also a good approach.

Dataflow-driven testing focuses on problems in component interfaces. A data element receives a value in one place and uses it in another place. Sometimes the value is misinterpreted by the receiving element. A good tool in dataflow testing is a CRUD (Create,

Read, Update, Delete) table, which defines who has what rights to the data element. Dataflow testing can also be based on flow diagrams. Both the CRUD table and flow diagrams make it easier to understand how the software is supposed to function. In addition, **control-flow-driven** testing can bring additional viewpoints to test planning that might otherwise be forgotten.

Message Sequence Charts are used in component design specifications for the S60 platform. These same charts are potentially valuable in understanding the client – server architecture better.

Extreme programming, despite the programming alternatives in it, has proved to be an excellent programming procedure that also includes testing. Another name for it is the buddy system, because not a single line of code of a complex component is written by just one programmer. The person sitting next to the programmer is constantly reading the code and is surprisingly able to discover most of the faults in it immediately.

Code review has for years been one of the most efficient defect finding techniques. Because it is a very slow technique, it should be focused only on the most risky components. The average progress of a code review is around 100 lines of code per hour. Unfortunately, this technique is not followed as often as it should be; because many people find it a very boring activity.

Static analysis utilizes tools to check the program. Nowadays, tools can check at least the following things:

- operations: write-read-write-read
- errors in CALL statements
- code impossible to execute
- risky constructs (such as pointers)
- unused variables

What can be quite frustrating sometimes is the fact that only very few of all the warnings received are really worth further investigation.

9.3 Testing Phases

The most common testing phases are introduced one at a time in this section with some tips on how to execute them. Though each test phase has its own special nature, they have many similarities. The phases covered in this book follow the V-model.

9.3.1 Documentation Testing

Program documentation plays an essential role in the success of testing. Documentation is often used as the only input when the testing activities are planned. Therefore the quality of documentation has a direct impact on the success of testing. The documentation can be improved in many ways. The following activities help in creating good-quality documents:

- **Use of a template** helps to keep all documentation consistent. Once the tester knows how to read the feature specification and where to find all information he or she can plan tests more easily and faster. The entire program personnel must naturally be trained in template usage, so that people know into which template to use.
- **Maintaining documents in one place**, where the tester can always find the latest version. In this way the risk of using an outdated version of a specification is removed.
- **Using pictures instead of words**, especially in User Interface (UI) design, is very much recommended. One picture says more than a thousand words.
- **A formal review process** helps the program to discover faults in the documentation before they are actually coded into the software. This is because, if the implementation has followed a complicated architectural design, it can turn out to be impossible to make any further corrections in the code. In the other words, defects that originate from the architectural design phase can be very expensive to correct. All architectural documentations should go through definitive review and approval processes.

9.3.2 Module Testing

A module is a set of programs that serves a predefined purpose within the entire system and is always owned by one single programmer. What this predefined purpose is in practice can vary a lot in different programs. It can be a single class for instance or a single

Dynamic Linked Library (dll) file. A bottom-up testing approach means that each component is first tested in isolation.

Module testing is very often given a name such as unit testing or component testing. In all cases it aims to discover how a particular module/unit/component is working compared to its specification. The module testing procedure is to write test scripts for all functions and methods so that, whenever a change causes a regression, it can be quickly identified and fixed.

The goal of unit testing is to isolate each part of the program and show that the individual parts are correct (Figure 9-6). Unit testing provides a strict, written contract that the piece of code must satisfy.

According to Wikipedia,[5] properly accomplished module testing affords several benefits.

- **Facilitation of changes.** Once the module test cases are created, the programmer can start using them. It is very likely that the module needs to be changed after the first test rounds. Good-quality module test cases encourage the programmer to change the implementation if needed, as re-running tests is very simple. A good set of module test cases covers the entire module; in other words, every line of code is executed during testing.
- **Documentation**. Module testing provides a sort of 'living document'. Clients and other developers aiming to learn how to use the module can look at the module tests to determine how to use the module to fit their needs and gain a basic understanding of its API and services. Module test cases embody characteristics that are critical to the success of the module. These characteristics can indicate appropriate and inappropriate use of a module, as well as negative behaviours that are to be trapped by the module.

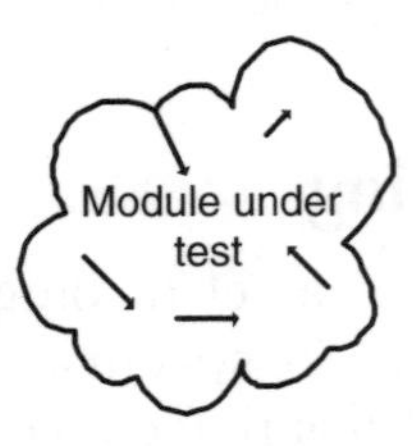

Figure 9-6. Module testing.

- **Simplification of integration**. By testing the parts of a program first and then testing the sum of its parts, integration testing becomes much easier.
- **Separation of interface from implementation**. Since the module testing is only aimed at verifying the module's internal behaviour, it is vital to understand the difference between internal and external interfaces. Wikipedia explains this in the following way:

  A common example of this is classes that depend on a database: in order to test the class, the tester often writes code that interacts with the database. This is a mistake, because a unit test should never go outside of its own class boundary.

Module testing only tests the functionality of units separately. It cannot provide any information on integration, performance or feature-level defects. Since its success depends 100 per cent on the quality of the documentation and the programmers' competencies, its result varies from program to program.

There are several good framework tools in today's world to speed up module testing. Some of these tools are textual and some graphical: Textual means that the test cases are written with, for example, NotepadOne and run in MS-DOS, whereas graphical means that there is a graphical dialogue and some graphical progress indicator. One such tool is JUnit, which is intended for Java-coded modules. JUnit is a freeware tool available over the Internet. Use of JUnit requires Java coding knowledge plus module testing and architecture knowledge. It is introduced briefly below.

Writing test code in JUnit involves the following:

- Create an instance of Test Case.
- Override the method runTest().
- When you want to check a value, call assert() and pass a Boolean value = true if the test succeeds.

Running two or more tests that operate on the same or similar sets of objects involves the following:

- Create a subclass of Test Case.
- Add an instance variable for each part of the fixture.

- Override setUp() to initialize the variables (e.g. establish a network connection).
- Override tearDown() to release any permanent resources you allocated in setUp.

Running several tests at once involves the following:

- Create a test suite OR.
- Let JUnit extract a suite from a Test Case.
- Run JUnit tests and collect test results.
- Make your suite accessible to a TestRunner tool with a static method *suite* that returns a test suite.
- The graphical user interface presents a window with:
    - a field to type in the name of a class with a suite method
    - a Run button to start the test
    - a progress indicator that turns from red to green in the case of a failed test
    - a list of failed tests
- The textual Test Runner shows the results on the system console.

9.3.3 Integration Testing in the Small

The ISEB standard differentiates integration testing between modules and integration testing between systems. The former is called integration testing in the small and the latter integration testing in the large. Sometimes this activity is called integration and testing and abbreviated as I&T. Integration testing takes as its input modules that have been checked during module testing, groups them intp larger aggregates, applies tests defined in an integration test plan to those aggregates and delivers as output test results the possible holes in the integrated system (Figure 9-7).

The different types of integration testing are Big Bang, Top Down, Bottom Up and Back bone:

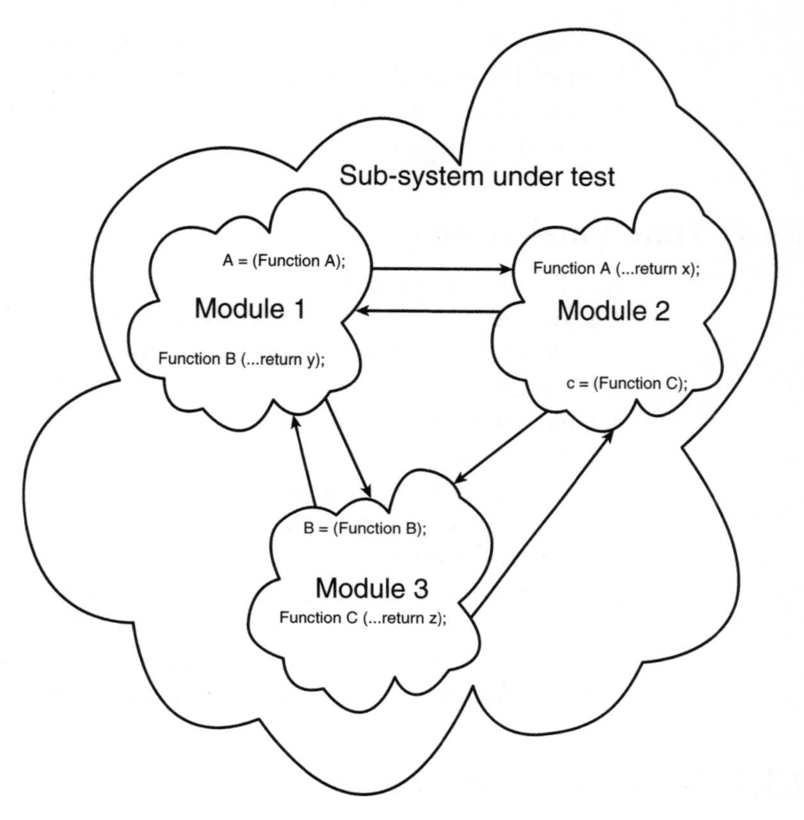

Figure 9-7. Integration testing in the small.

- **Big Bang** means that the entire pool of modules is integrated at one time and integration testing is done on the whole system.
- **Top Down** practice equates to a situation in which the modules that architecturally make up the lowest level of the system are integrated and tested first, whereas **bottom up** constructs the system in the opposite order.
- **Back Bone** means that the modules that are used most are combined first; such a sub-system is called a system back bone. This back-bone sub-system is tested first and after that other modules and sub-systems are integrated within it.

Where module testing is done by programmers, integration testing is often accomplished by a separate testing team specialized in system integration defect discovery.

There are also tools available for integration testing purposes. Most of the module testing tools and frameworks are also suitable for integration testing purposes. The usage of such tools is very common nowadays.

9.3.4 Functional Testing

Functional testing covers how well the system executes the functions it is supposed to execute – including user commands, data manipulation, searches and business processes, user screens and integrations. Functional testing also covers the obvious surface type of function, as well as the back-end operations (such as security and how upgrades affect the system).

In some product programs functional testing may capitalize on using functional testing tools and frameworks. Sometimes finding a suitable tool is difficult because of the special nature of the product. For example, testing the functionality of a smartphone is very different from testing the functionality of domestic appliances.

Functional testing is very practical task to carry out. It is also easily understandable by people who are used to similar kinds of product.

9.3.5 Non-functional Testing

Non-functional testing aims to find defects in product performance, stability and other things that are not measurable in terms of functional correctness. Performance testing can be carried out in many ways, some of them being as follows:

- **The maximum load the product can handle.** For this one needs to define the selection of the stimulus to be used to load the product as well as the maximum acceptable response time. This is also called a product's **high watermark** (HWa) definition.
- **The performance of the product under high load for a longer time.** For this one needs to define how what percentage of the HWa is to be used and for how long a time. The duration varies a lot from one product to another. Sometimes the load is changed during the load period, as shown in Figure 9-8.
- **A combination of several very time-critical user actions with a predefined load**. This approach is very important in testing products with strict response requirements.

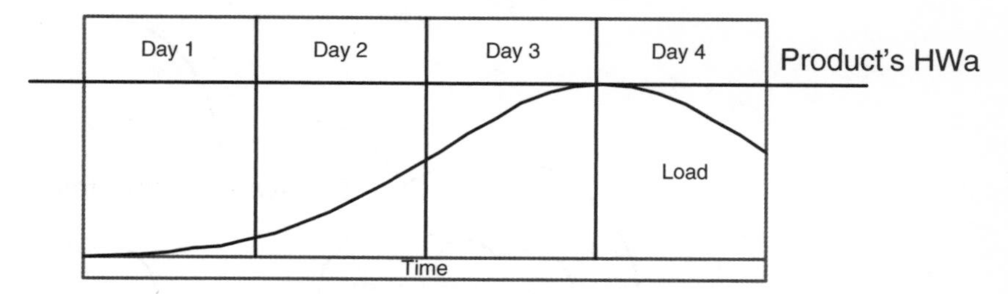

Figure 9-8. Example of a performance test load.

One testable item not related to functional correctness is, for example, security. Product security can be measured in numerous ways. A few can be explained as follows:

- **Loss of confidentiality of information** means that the product erases or allows some other entity to erase confidential information or data.
- **Compromise of integrity of information** means that the product allows modification of information or data in a wrong way either by itself or by some other entity.
- **Denial of Service (DoS)** verifies whether the product becomes jammed under a high enough load. This is an important test in smartphones as sooner or later hackers will try to harm mobile phone users in one way or another.
- **Misuse of service, systems or information** means that the product allows an unauthorized entity or application access to confidential data or information.

9.3.6 Integration Testing in the Large

When larger entities are combined, it is necessary to verify how they work together. Within one system or product that activity is called Integration testing in the small (ITS). Among several products the activity is called integration testing in the large (ITL).

In software, we are normally concerned with integration at two levels. First there is the integration of components at the module level into a system – sometimes known as component integration testing or integration in the small. Second there is the integration of systems into a larger system – sometimes known as system integration testing or integration testing in the large.

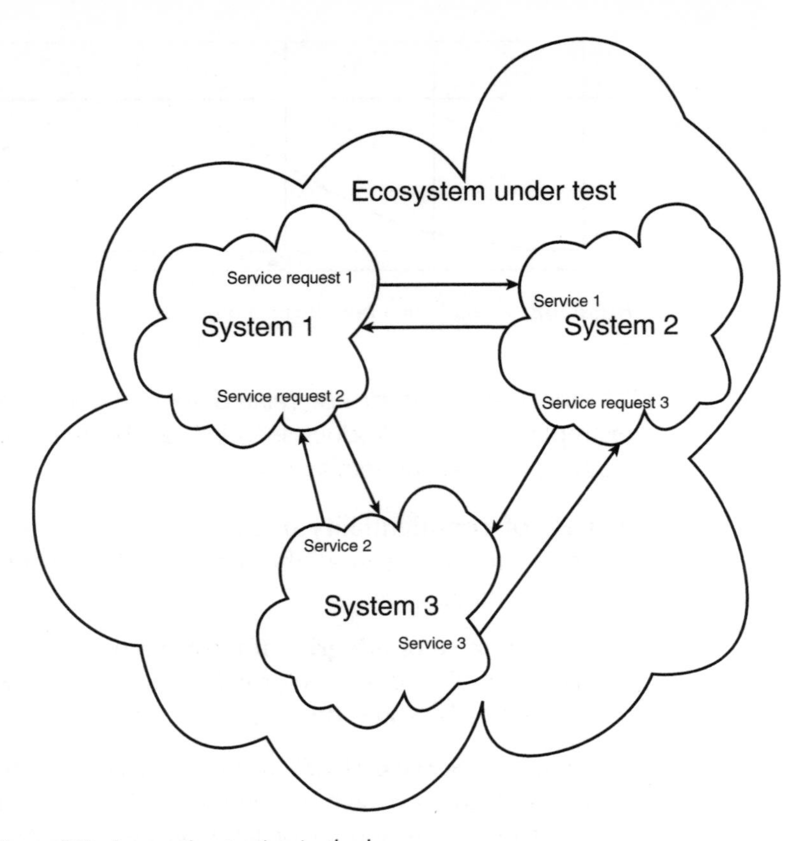

Figure 9-9. Integration testing in the large.

In a mobile ecosystem environment ITL verifies how a device functions with other devices and different networks and servers. This is illustrated in Figure 9-9.

Common standards help in implementing compatible products, but, especially in the software world, this is easier said than done.

9.3.7 The Real User Experience (TRUE)

As mentioned earlier in this book, the only way of discovering all defects in a product is long-term use by many users. Unfortunately this can only be achieved after the product is on the market. Since this kind of feedback would be very valuable to the program, user experience is often simulated.

Testing real user experience (TRUE) is normally carried out as soon as the product can be used in a meaningful manner, in other words, once the product has enough functionality to be used by anyone. It provides feedback from real-life usage during the R&D phase, when fixing is still possible.

How many users are required to bring enough information to the program? For example, in a mobile phone program there are normally around 150 to 300 users selected.

TRUE testing does not need any test cases because a selected set of people from a target consumer group are using the product prototype in the way they want to. These people are trained to report all defects they discover in the product.

There are certain preparations that the entire product program and its test manager need to do or have before TRUE testing can start:

- global system specification
- clearly defined development requirements
- controlled system maintenance
- global control via TRUE central enabling visibility of test user resources
- visibility of networks used (network elements and features supported)
- understanding TRUE tester preferences
- common support process
- common reporting process
- very early batch feedback to confirm that TRUE test failure statistics correlate with field feedback results
- fault symptom codes easy to use and able to correlate with field feedback
- TRUE testing maturity checklist for TRUE ramp-up
- measured flashing support to maintain quality of service
- software version always maintained according to the actual phone state
- TRUE user's profile to include operator system capabilities
- defined process for sim-card provisioning to maintain needed features

- use regional requirements to define TRUE priorities
- minimum level of TRUE users defined

TRUE testers need to be trained in numerous things such as:

- the quality of the TRUE reporting system
- program commitment to planning and analysis of TRUE test output
- use of mobile functionality in varied bearer networks
- early batch feedback through extended TRUE testing
- market visibility through extended presence and scope of the test users
- special focus tests to prioritize usage of the most critical applications
- standard statistics defined as program measures and quality metrics (understanding of what programs actually need)
- central support for data delivery, analysis and comment
- quality reporting through defined feedback channels
- structured reporting criteria, tailored to meet program needs
- clear entry criteria for programs entering TRUE testing
- management commitment to single feedback solution
- commitment to program and support resources
- high volume of active TRUE testers
- structured software flashing support
- clear roadmap for planning TRUE test support per site
- clear understanding of TRUE users and operator system capabilities
- effective sim-card provisioning across operators, updated with the latest features
- product prioritisation based on regional requirements
- global visibility and control of available TRUE test users

9.4 What Then?

Once testing discovers a defect, it is automatically assumed to be the result of an error in the product. However, testing can also contain defects; especially if people other than designers and programmers carry out the testing. The tester may misunderstand the specifications or make assumptions about a product's behaviour and therefore report a defect that does not really exist at all. Testers should always follow the commonly agreed rules and procedures when planning, executing and reporting testing activities.

After the very final test round, when the product has proved to have reached commercial quality, it is deployed. The maintenance requirements are decided and tailored into each program. Figure 9-10

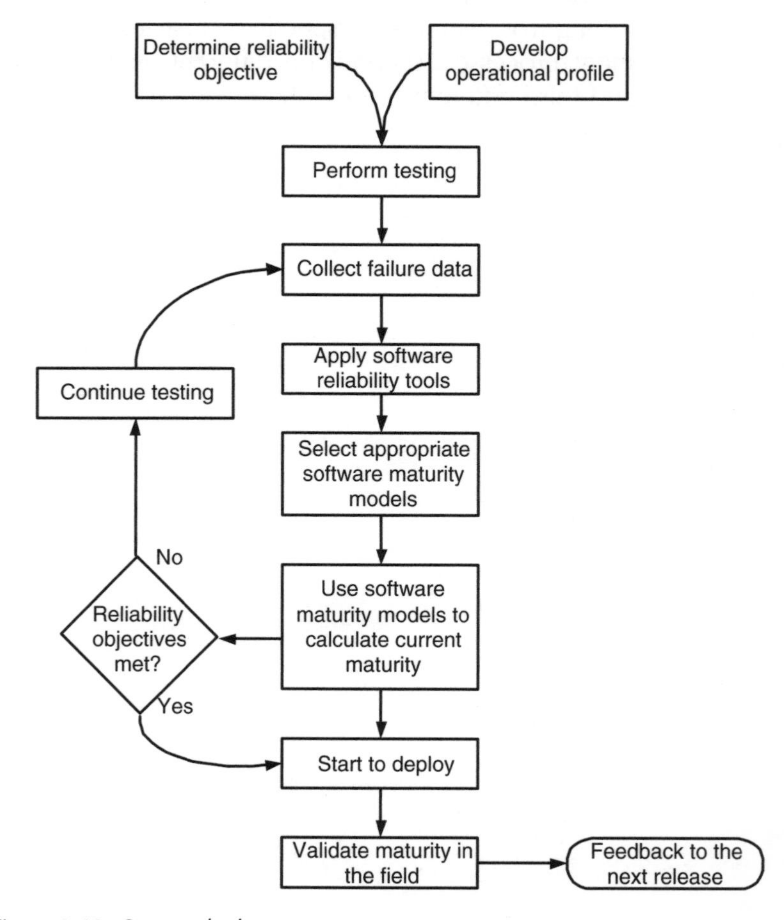

Figure 9-10. System deployment process.

shows at a high level the actions needed for deciding the requirements for maintenance work.

More information on defect handling and management is given in chapter 11.

9.5 Summary

Testing that is done professionally is the only way to discover defects in a system. In the history of software testing was introduced as an equal activity to coding not too long ago, and negative attitudes can still be seen in various projects and organizations. However, most projects are ready to pronounce testing as being an undisputable activity for finding the holes in a product. This chapter has discussed testing from various perspectives, first testing in different development processes; second the different testing techniques and tools and third different testing phases one by one.

Chapter 10: The Testing Environment

Smartphone testing involves quite extensive requirements, in terms of both money and competence, on the equipment and environment needed. In the earlier testing phases, such as module and integration testing, the requirements are mostly competence-related ones such as knowledge of module testing tools and scripts. The later the testing phase is, there is no question that more money needs to be invested in external tools and other elements. Figure 10-1 shows an example of the required elements in an average phone program as far as test equipment and competencies in each testing phase are concerned. For example, the minimum requirement to carry out good and extensive interoperability testing is that there is access to all needed servers and network elements. If the program has administrator access to these elements, so much the better. Owning such a network and servers is very expensive and not vital since all tests can be executed over a publicly available network. The question is how difficult troubleshooting will be without access to the network logs.

S60 Smartphone Quality Assurance Saila Laitinen

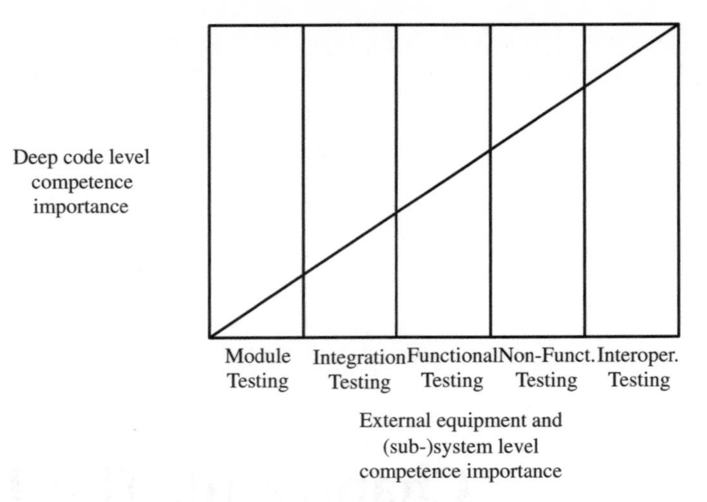

Figure 10-1. Phone program needs for equipment and competencies in different test phases.

The testing environment can be divided into two:

- **External equipment.** Over-the-air (OTA) testing is enabled with different servers. Different tools are required to help, for example, in test execution automation and test result analysis.
- **Test data.** Data is needed to speed up test case execution both in FUTE and NOFTE.

Both of these elements are introduced in this chapter for each test activity.

10.1 Module Testing

Module testing has an indisputable and genuine role in every software program. It is one of the most important testing activities and it normally has the cheapest requirements for the test equipment required. It is also the cheapest testing activity as the defects found during module testing are very cheap to fix. The S60 customer program's module testing does not set any extraordinary requirements, such as servers and test tools, for the external equipment. Of course, some external test automation tools can be used to speed up the test execution, but, since the module testing are always based on pre-written test code that can be run whenever needed, the added value of a test automation tool or framework can be very little.

The biggest investment required in carrying out the most efficient module testing in the S60 customer program is to make sure that the programmers have the required knowledge of Symbian, S60 and the module testing techniques introduced in chapter 9. In addition to above, programmers should also have a clear understanding of what kinds of test classes are available in a platform release as this may help the implementation of the required test code.

10.2 Integration Testing in the Small

Since Integration Testing in the Small (ITS) is aimed at finding defects on a module's APIs, it is one step more complex than module testing. ITS focuses tests on the 'external' APIs and services of components. As mentioned in chapter 2, the platform is implemented either one feature or some features at a time.

As indicated in Figure 10-2, sometimes the implementation orders of platform and customer APIs do not accommodate each other. This can cause a situation where the customer program is ready to start testing in areas where there is either no implementation at all or there is only half-ready code of a counterpart on the platform side that can be used. In such cases the customer program needs either to delay the testing activities until the platform has delivered mature enough components or to build their own stubs and/or drivers to be used to replace the missing APIs.

10.3 Functional Testing

As stated in chapter 1, functional testing starts to have relatively large financial requirements for the program. The external equipment needed to run all functional tests delivered with the 2.x-platform deliveries is listed below. In those cases where a GSM network with General Packet Radio Service (GPRS) access is required, such access is a minimum requirement to enable execution of the test. If administrative access can also be obtained, possible troubleshooting of those cases that fail is possible. Sometimes this can be extremely valuable, especially if there is a question whether the defect is on the network side and not in the terminal.

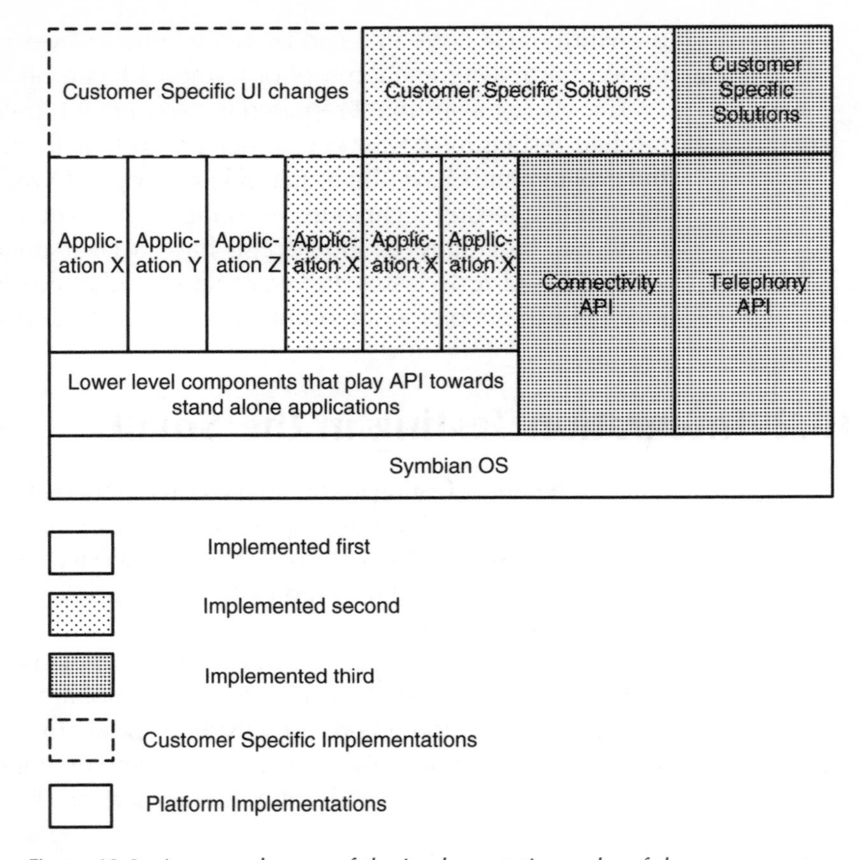

Figure 10-2. An example case of the implementation order of the components.

10.3.1 Common

The required elements (both environment and data) for the functional testing of common parts of the S60-based phone are introduced in this section.

Memory Card Application testing needs a Windows environment, terminals (prototypes), several of the chosen type of MMC cards with at least one corrupted one to test exceptions and a MMC card reader. On the data side, some files to store to the MMC cards are also needed.

Enabling MMC Hot Swap testing needs a Windows environment, terminals, a MMC card and a MMC card reader.

The **Application Installer** testing needs a Windows environment, a terminal SIM card supporting Java MIDlet downloading and a server to download Java MIDlets. As data the Java Midlets are needed in testing.

Device Manager testing needs a Windows environment, terminal prototypes, a Device Manager (DM) server and a Nokia Terminal Management Server (NTMS) with verification page. This server needs to support all following: WAP Push using SMS bearer, HTTP 1.1, Secured HTTP SSL v. 3.0 and TLS 1.0, OMA Provisioning (OTA) and SyncML version 1.1.

General Settings testing needs a Windows environment, terminal prototypes, a SIM card that supports alternate line service, a SIM card that does not support alternate line services, a Network Identity and Time Zone (NITZ) and OMA provisioning. As data some animated, corrupted and large gif images are necessary.

Application Shell testing needs terminal prototypes and the chosen type of MMC cards. As data also some applications for installing on the MMC card are necessary.

Context sensitive Help testing requires terminal prototypes.

Offline mode testing needs terminal prototypes, at least a full coverage of GSM networks with GPRS, MMSC and Email servers, as well as IR and BT capable devices.

Location application testing needs terminal prototypes, as well as a network that supports the location service.

10.3.2 UI Customization and Personalization

The functional testing of elements (both environmental and data) related to UI customization of the S60-based phone are introduced in this section.

Profiles application needs terminal prototypes, ALS, MMC and coverage of a GSM Network. As data, different tone formats are needed. These exist by default in the device, can be created by the user with a voice recorder, can be received as Smart messages and saved to phone, can be received as mail attachments and saved to the phone, can be received as MMS messages and saved to the phone or can be transferred to the phone via PC connectivity.

UI themes testing needs terminal prototypes and, as data, different kind of images and themes are need to run the tests.

Personalization applications testing needs a Winsows environment, terminal prototypes and the generated images and implemented themes as test data.

Testing of **preset download folders in browser bookmarks** needs terminal prototypes, a SIM card with network connection enabled Circuit Switched Data (CSD), High Speed Circuit Switched Data (HSCSD), GPRS access points, GSM network and dialling servers. As data the bookmarks are needed.

Testing of **embedded download links in applications** needs terminal prototypes.

Pinboard testing needs terminal prototypes and the chosen type of MMC cards. As data, some images, a notepad memo, voice recorder files and WAP bookmarks, as well as saved WML cards, are needed.

10.3.3 Local Connectivity

The functional testing elements (both environment and data) of the S60-based phone local connectivity are introduced in this section. Naturally, if the device does not contain, for example, a Bluetooth port, the BT-related aspects do not apply in the program.

Bluetooth connectivity testing needs terminal prototypes as well as BT-enabled devices. As data, some files to be transferred via BT are needed.

Infrared connectivity testing needs terminal prototypes as well as IrDA-supported devices. As data, some files to be transferred via IR post are needed.

Universal Serial Bus (USB) connectivity testing needs terminal prototypes and a USB cable, as well as other USB devices. As data, some files to be transferred via USB are needed.

10.3.4 Networking and Data Bearers

The functional testing elements (both environmental and data) of networking and data bearers of the S60-based phone are introduced in this section.

HTTP Protocol testing needs terminal prototypes and a World Wide Web server. As data, some HTML and XHTML test material is needed, as well as HTTP error codes.

WAP Protocol testing needs terminal prototypes, a WAP server and, as data, some material that supports WAP.

Testing of **GSM Circuit Switched Data** needs terminal prototypes and a GSM network with CSD capability.

Testing of **GSM High-Speed Circuit Switched Data** needs terminal prototypes and a GSM network with HSCSD capability.

General Packet Radio Service (GPRS) testing needs terminal prototypes, GSM network coverage with GPRS capabilities, two other terminals with GPRS capabilities and three SIM cards (one with a GPRS subscription, one without a GPRS subscription and one with a static IP address), as well as a PC with Email, Web browser, File Transfer Protocol (FTP) and modem driver for phone and fax.

Testing of **Enhanced Data Rates for Global Evolution (EDGE)** needs terminal prototypes, EDGE network coverage, two other terminals with EDGE capabilities, the latest S60 phone software available and three SIM cards (one with an EGPRS subscription, one without an EGPRS subscription and one with a static IP address), as well as a PC with Email, Web browser, FTP and modem driver for phone.

Connection manager testing needs terminal prototypes, two SIM cards (one with multiple PDP contexts, MMS, GPRS and CSD enabled and one with no support for multiple PSP context) and a laptop with Bluetooth and IRDA settings. As data, the settings need to be in place prior to testing.

10.3.5 Telephony

The functional testing elements (both environmental and data) of the S60-based phone telephony-related features are introduced in this section.

Telephony testing need terminal prototypes, a GSM network (with support for conference calls, call charging, call transfer and alternate line service (ALS)), one blocked SIM card (with the PIN code locked), one rejected SIM card (the PUK code is rejected), one unsubscribed SIM card (without connection to the network) and a clock for testing the duration between redial call attempts.

Testing of **Fax and Data calls** needs GSM network coverage (with CSD capability), terminal prototypes, a laptop with BT, IrDA and ProComm fax, two other terminals with the capability of receiving data calls, two other devices with the capability of sending and receiving faxes and a SIM card that supports waiting data and fax calls. As data, the fax content is needed.

Logs application testing needs terminal prototypes, two other S60 terminals, two different SIM cards (one that supports ALS and one that supports CLIR) and a stopwatch. As data, SMS messages (incoming and outgoing SMS messages, delivered and pending SMS messages, SMS messages with failures), data calls, fax calls and voice calls without a number, with a private number and generated by SIM ATK are needed.

General Log testing needs terminal prototypes, two different SIM cards (one with support for ALS and one with support for CLIR). As data, SMS messages (incoming and outgoing SMS messages, delivered and pending SMS messages, SMS messages with failures), data calls, fax calls and voice calls without a number, with a private number and generated by SIM ATK are needed.

10.3.6 Multimedia

The functional testing elements (both environmental and data) of the S60-based phone multimedia features are introduced in this section.

Camcorder application testing needs terminal prototypes or emulator environments.

Image Viewer testing needs terminal prototypes, a GSM network with MMSC, Email servers, IR and BT capable devices and the chosen type of MMC card.

Media Player testing needs terminal prototypes, a GSM network (with the capability for streaming over EGPRS, GPRS and HSCSD bearers), MMSC, Email servers, an IR- and BT-capable device, the chosen types of MMC cards (normal and locked), a headset, a stereo headset, a Bluetooth headset, SIM cards with (E)GPRS and HSCSD support, a PC as a streaming server, content creation tool(s) and a USB cable. As data, a wide range of supported and unsupported files for both local and streaming playback (supported file formats for video include: 3gp, MP4 and RM – note: only MP4 files containing supported codecs can be played; supported file formats for audio include: AAC, AMR, AU, AWB, MID, MP3 and WAV) and streaming links (RTSP URLs) stored in RAM files.

Media Gallery testing needs terminal prototypes, a GSM Network with MMSC, Email servers, an IR- and BT-capable device and the chosen type of MMC cards (normal, locked, corrupted and read-only). As data, some images (in, for example, the following formats: jpeg, gif 87 & 89a, png, tiff/f, mbm, bmp, wbmp, Smart Messaging OTA Bitmap – GMS pictures –wmf, exif and ico), some audio files (in, for example,. the following formats: rng, wav, au, amr, awb, midi, mp3 and aac), some video clips (in, for example, the following formats: 3gp and rm), a streaming link (ram file), some non-media file types (for example, HTML, text, EXE and DLL), files that are too large in size, video clip in NIM format and some media files that are application-specific/proprietary.

Voice Recorder testing needs terminal prototypes, a GSM network, MMSC and Email servers, as well as an IrDA and BT device. As data, some contacts need to be created in the prototype.

10.3.7 Personal Information Management (PIM)

The functional testing elements (both environmental and data) of the S60-based phone PIM features are introduced in this section.

Contacts testing needs terminal prototypes, several SIM cards with different maximum sizes for memory entities (one SIM card with full memory), an IrDA device to receive contact cards, a Bluetooth device to receive contact cards, another S60 or GSM phone to receive SMS messages and SIM cards with and without service numbers.

Calendar testing needs terminal prototypes, a GSM Network, MMSC, Email servers and IrDA- and BT-capable devices.

Notes testing needs terminal prototypes, a GSM network, MMSC and Email server.

Clock testing needs terminal prototypes and a GSM network that supports NITZ.

File Manager testing needs terminal prototypes, a GSM network, MMSC, Email server, IrDA and BT devices and the chosen types of MMC card. As data, some images (in, for example, the following formats: jpeg, gif, bmp, wbmp and tiff), a GSM picture file, a link file, a sound file (that can be played with MediaPlayer), a play list, a ringing tone file, a sis file, a video file, one corrupted file and one unsupported file are needed as well.

Remote Synchronization testing needs terminal prototypes, a GSM network, a remote sync server and a PC.

10.3.8 Messaging

The functional testing elements (both environmental and data) of the S60-based phone messaging features are introduced in this section.

Messaging center testing needs terminal prototypes, GSM network coverage and MMSC and Email servers. As data, access rights to a remote mailbox need to be in place for at least one server.

Short Messaging (SMS) testing needs terminal prototypes and GSM network coverage. As data some test messages are also needed.

Smart Messaging testing needs terminal prototypes, GSM network coverage with GPRS, MMSC and Email servers. The chosen type of MMC card, a hands-free device and IrDA and BT devices.

Multimedia messaging testing needs terminal prototypes, GSM network coverage with GPRS, MMSC and Email servers, the chosen type of MMC card, a hands-free device and IrDA and BT devices. As data, some images (in both supported formats, such as jpeg, gif, png and wbmp and unsupported formats), images with the following sizes: 640 × 480 pixels and 160 × 120 pixels, audio files in the AMR format, video clips (in the following formats: 3gp, nim and mp4 as well as some other audio formats) and a SIS package of the XAMPLE4 testing tool are needed.

Email testing needs terminal prototypes, GSM network coverage, an Email server, GPRS and HSCSD access points and four different remote mailboxes (POP3, POP3-SSL/TLS, IMAP4 and IMAP4-SSL/TSL). As data some pictures and other attachment file types are needed.

Cell Broadcast testing needs terminal prototypes and GSM network coverage that supports cell broadcasting.

Testing **receiving and sending messages via IrDA and BT** needs terminal prototypes, a GSM network and other IrDA and BT devices. As data, some messages are needed.

Testing of **OMA Instant Messaging** needs terminal prototypes, GSM network coverage with the IM service availability, the chosen type of MMC card and another IM device.

OMA Presence Server testing needs terminal prototypes, GSM network coverage with the Presence Service (PS) availability, the chosen type of MMC card and another PS device.

Presence application testing needs terminal prototypes, GSM network coverage with PS availability and MMSC. As data, MyLogo files of different sizes and some wireless village user IDs are needed.

Presence API testing needs terminal prototypes and GSM network coverage with PS availability.

Testing of **OMA client provisioning** needs terminal prototypes, GSM network coverage with OMA client provisioning service and a SIM card.

10.3.9 Browsing

The functional testing elements (environmental and data) of the S60-based phone browsing features are introduced in this section.

Browser testing needs terminal prototypes, GSM network coverage with GPRS, CSD, HSCSD capabilities, another S60 phone, a SIM card with only GSM Voice Call service, a SIM card with only GSM data service, a SIM card with HSCSD data service, a SIM card without HSCSD data service, a SIM card with GPRS data service and a SIM card without GPRS data service. As data, XHTML, I-Mode and HTML pages, background images and WAP push messages are needed.

Security testing needs terminal prototypes, GSM network coverage and several SIM cards with the known PIN, PIN2, PUK, PUK2 codes, security codes and pre-programmed master code.

Digital rights management testing needs terminal prototypes, GSM network coverage, the Nokia content publishing toolkit, another S60 phone or another phone that supports DRM, two SIM cards from different operators with all possible data services support, a website for the content, an MMS centre, a Multimedia sender tool and a WAP push tool. As data, several different types of DRM messages are needed (for example, gif, animated and not animated, jpeg,, progressive and sequential, png, midi Audio, sp-midi, amr and amr-wb).

Fax testing needs terminal prototypes, GSM network coverage, a fax device, two other terminals with fax support, a PC with Windows NT/2000 OS, a connectivity pack (BT, ProComm and Winfax Pro) and at least two SIM cards for different operators. As data, a three-page test sheet is needed.

Testing of **Synchronized Multimedia Integration Language (SMIL)** needs terminal prototypes, GSM network coverage, a SIM card that contains Service Dialling Numbers (SDN) storage, a SIM card that does not contain SDN storage and a SIM card that contains empty SDN storage.

Service Dialling Numbers (SDN) testing needs terminal prototypes, a GSM network, a SIM card that does not contain SDN storage and a SIM card that contains empty SDN storage.

10.4 Performance Testing

As explained in chapter 9, performance testing has many aspects. The maximum load a product can handle is one measurement of performance testing; another is how long the product can function under a certain load. Comprehensive performance testing was not part of too many phone programs until the early 1990s. Its role has become more important in connection with the latest phone models that contain more and more features.

The extent of performance testing of a S60-based phone is purely a program's decision. Figure 10-3 shows some of the common elements needed in doing basic performance testing on a smartphone by using multimedia messages as the load.

A general load generator is a good tool for creating loads on the device in a controlled way. With a good load generator tool, the maximum load a device can handle with the predefined response times can be defined. In product performance testing it is very important to include all modules in the System Under Test (SUT) as the default is that all modules can contain a performance defect and that defect can cause failures in the overall functionality of the product. This means that the load generator should be able to transfer the load over the air to the phone. This load can be, for example,

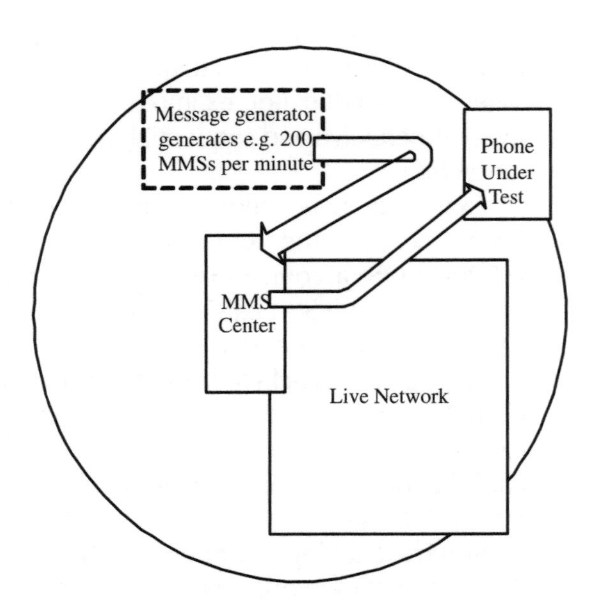

Figure 10-3. MMS message generator example.

a combination of short messages, multimedia messages, smart messages and different voice calls (both incoming and outgoing), as well as other connectivity traffic.

10.5 Interoperability Testing

Smartphone compatibility with the external world is very crucial for its success. Therefore correctly performed interoperability testing is highly recommended, even though it can be relatively expensive.

Table 10-1 describes at a high level what network elements are needed for the execution of interoperability tests on a smartphone. Interoperability is a condition achieved among elements of the Open Mobile Alliance (OMA) System Architecture when services and content can be exchanged directly and satisfactorily between them. Figure 10-4 introduces the elements needed for Multimedia messaging IOP testing on a S60-based phone.

Figure 10-4 shows clearly that to execute IOP tests for MMS functionality, the program needs access to several very expensive network elements. Very many publicly available networks already support multimedia messaging and therefore running the tests on their networks is, of course, possible, but, as mentioned earlier, the troubleshooting of failed cases is very difficult, if not impossible, without access to the log files for the network elements.

Nokia has its own IOP laboratory in Finland with the latest technologies and this laboratory can also be used for Licensee S60-based phone IOP testing activities if the fees and timing have been mutually agreed. The same laboratory not only has Nokia's own

FEATURE UNDER TEST (FUT)	NEEDED HARDWARE
Smart messaging	SMS Center
Instant messaging	IM group server including dynamic phonebook, terminal gateway and WAP gateway
Presence	Presence server and subscriber database
Voice calls	Network
Multimedia messaging	Multimedia messaging center and picture messaging center (e.g. NAMP)

Table 10-1. Hardware needed for IOP testing.

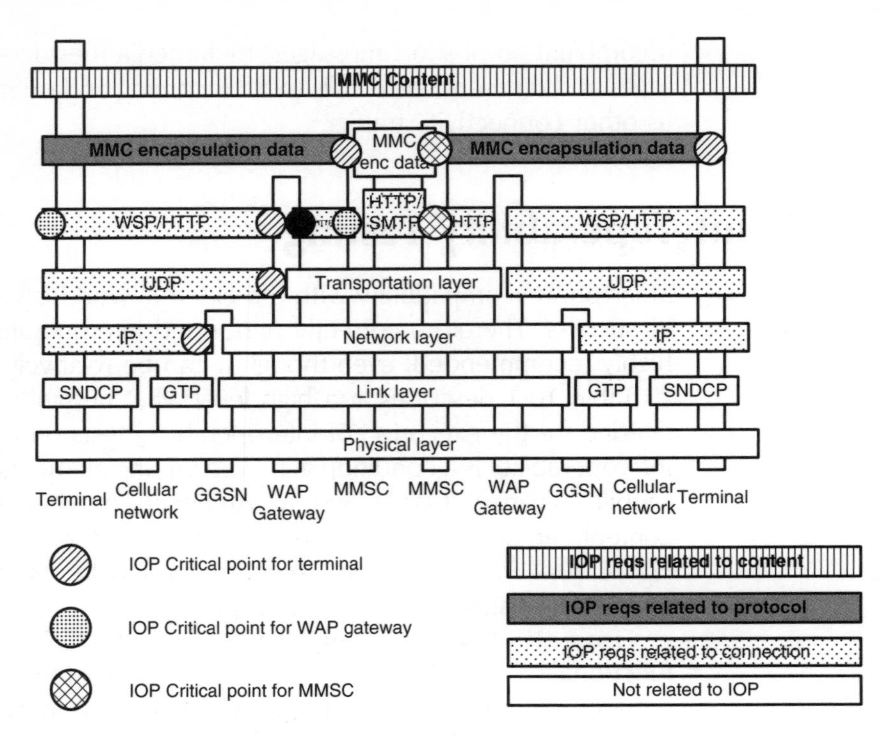

Figure 10-4. Example of S60-based phone MMS IOP elements.

network elements but also other vendors' servers, which provide very comprehensive IOP testing possibilities for customer programs. If agreed, the troubleshooting can also be done by Nokia IOP laboratory personnel.

10.6 Miscellaneous Testing Activities

A smartphone product program should also consider other testing activities in addition to thise discussed above. Some of these activities may have requirements related to the environment where the testing is done. One kind of miscellaneous testing activities is certification. Certification certainly sets some requirements on the external equipment needed and this equipment is introduced below.

10.6.1 Certification

The overall requirements of S60-based phone certification were explained in chapter 5. This chapter provides some information on what kind of environmental requirements there are in certification procedures.

One of the most significant environmental requirements in certification testing comes along with Java certification. Running a full certification test set, which covers all JSRs, requires as minimum one server equipped with Windows 2000 or XP Professional, terminal prototypes and network coverage. In practice, up to four servers and six prototypes reduce the time needed for testing. The other certification area that imposes additional requirements on the equipment needed is Bluetooth certification. However, since there are many companies around the world that provide Bluetooth certification testing as a service and they have all the necessary equipment in place, the terminal program should consider using one of these houses for the purpose.

10.6.2 Usability

Usability verification is also a very important testing activity in a smartphone program. It tries to find possible logic problems in the product's user interfaces that might cause end-user dissatisfaction in its usage. Although the S60 platform provides UI components that are already usability tested and verified, the customer program may still want to change the UI layout. The program needs to decide whether further usability testing is needed or not. If the program ends up deciding that they also need to carry out usability testing as one testing activity, they should also consider what kind of requirements are necessary in the testing environment.

The basic environmental requirements in doing proper usability tests are the following:

- Product prototype to be used by the tester:
    - an isolated room where the tester can use the product in piece and quiet
    - microphones and speakers for hearing the tester speaking and for making possible comments as well as for communicating with the tester
    - a room for observers

- Video recorder:
    - a recorder to record which buttons testers pushed while testing and when they pressed them

In addition to the above, appropriate data analysis tools are also needed for decompressing how the user used the phone.

10.7 Summary

A meaningful test environment is an absolute must for realizing testing activities in any product program. In a smartphone program this should be taken into consideration early on, because building a test environment is not an easy task. It requires proper network access, numerous server accesses, other devices and lots of test data and tools. This chapter has introduced those elements that have to be in place and those that would be benficial but are not absolutely mandatory to test a S60-based device.

Chapter 11: Defect Analysis

There are two things that can enable a product to achieve the necessary quality and freedom from error, if they are done right. One is the testing and the other one is sensible error fixing. Testing is intended to find defects in priority order and sensible fixing corrects them in priority order. However, these two priority orders can be different. In a product program, it may not always be wise to fix all the defects discovered, but instead handle them as known issues. This, of course, raises the question of why we did the testing to discover defects that we do not plan to fix. Sometimes defects in very complex components can be very safe and simple to fix. However, understanding what pieces of the functionality are so critical and important to the customer that they should be error free, no matter how big the risk of regression, is as important as recognizing which defects are very safe to fix.

At a certain stage in a program it can be worth fixing all discovered defects and, then, after that analysing which fixes are to be integrated and which not. This activity is called gatekeeping and the person responsible for it is called the gatekeeper. Figure 11-1 shows the sorts of issues a gatekeeper should consider for each fix.

S60 Smartphone Quality Assurance Saila Laitinen

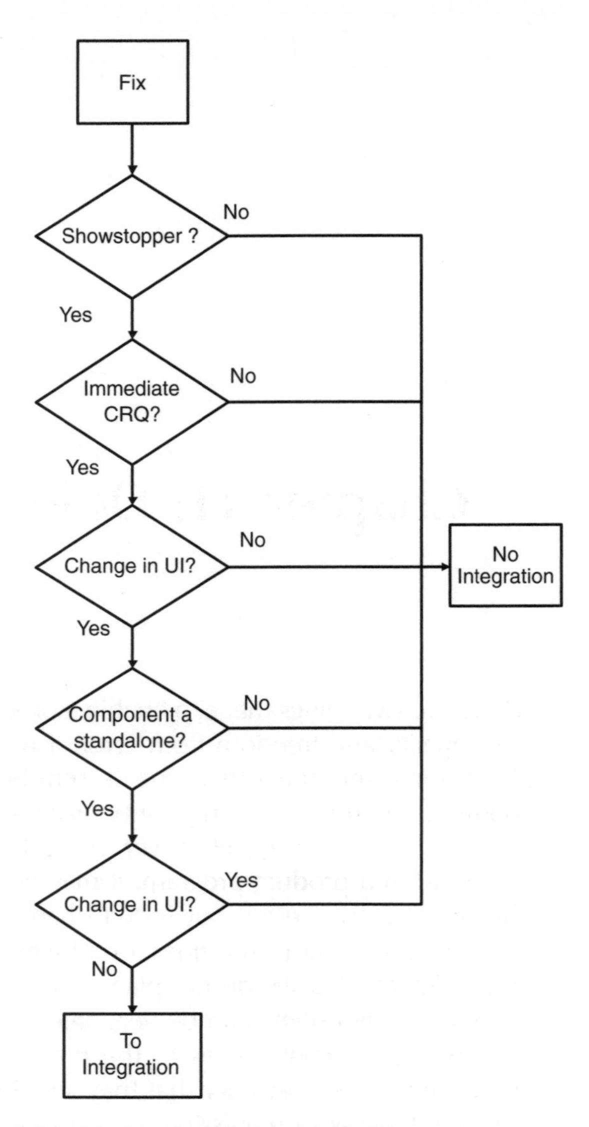

Figure 11-1. A simplified gatekeeping decision table.

Figure 11-2 shows how to prioritize the order of fixing. To simplify the elements of prioritizing, the product program should be clear about how the customer uses the product. In a smartphone, of course, a voice call remains one of the most used features in spite of all the newer features. And short messaging is another widely used functionality. These two features and related applications need to be very error-free or phone usage may lead to high dissatisfaction

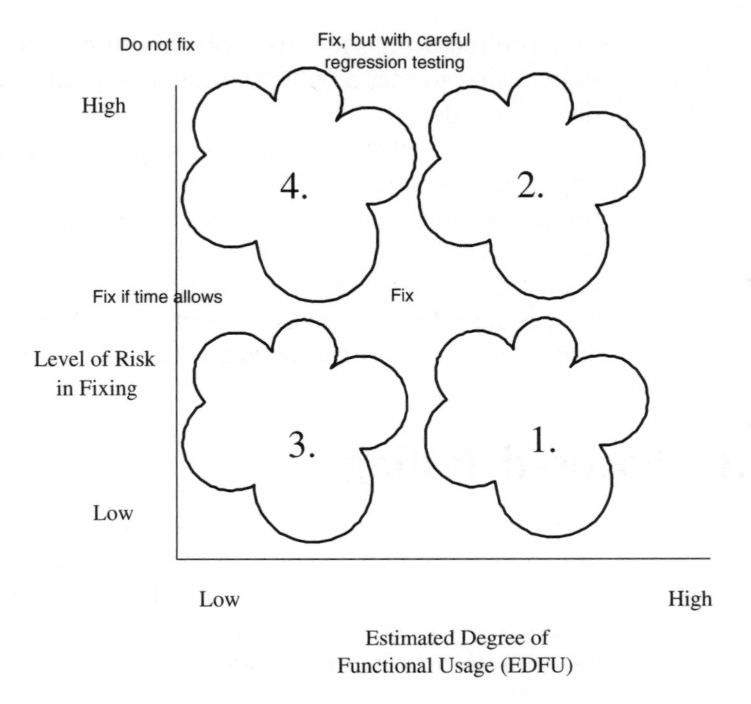

Figure 11-2. Fixing order.

and rebate rates, whereas the customer can ignore a misfunction in some feature or application that is not often needed.

The person in charge of deciding which defects are to be fixed first is called an error manager. The error manager needs to be in place before testing starts. He or she should follow the 'three-D-rule'; The words decide, define and distribute make up the 'three-D-rule':

- **Decide** means that the error manager should decide defect priority.
- **Define** means that the error manager needs to define characteristics and descriptions for each category (maybe even use simple and practical use-case examples).
- **Distribute** means that the error manager should share the information with all parties as well as monitor that the distribution rules are followed.

This section explains the importance of proper prioritizing activity on the known discovered defects and describes some tools to achieve this. The following sections explain both how to have the best possible testing so that a tight schedule is kept and what to do with the discovered defects.

One, very often forgotten, standpoint in defect analysis is the impact that a particular defect, if remaining unfixed, will have on the product's Field Failure Rate (FFR) and Mean Time Between Failures (MTBF). MTBF is a future value, which can be estimated based on known defects, testing coverage, product complexity and the use of the product. FFR is a measurement of product quality and reliability after shipping. It is the average time between customer-reported defects.

11.1 Focused Testing

Testing activities should be focused on areas that are commercially the most important to the company. Such areas are, for example, those pieces of functionality that hold the biggest estimated degree of functional usage (EDFU). Another viewpoint that can be used in focusing testing is to consider the most problematic and risky components.

Figure 11-3 shows one way to estimate defect criticality. The darker the area, the more critical the defect is. Naturally, the greater the EDFU, the more important it is for the component to work.

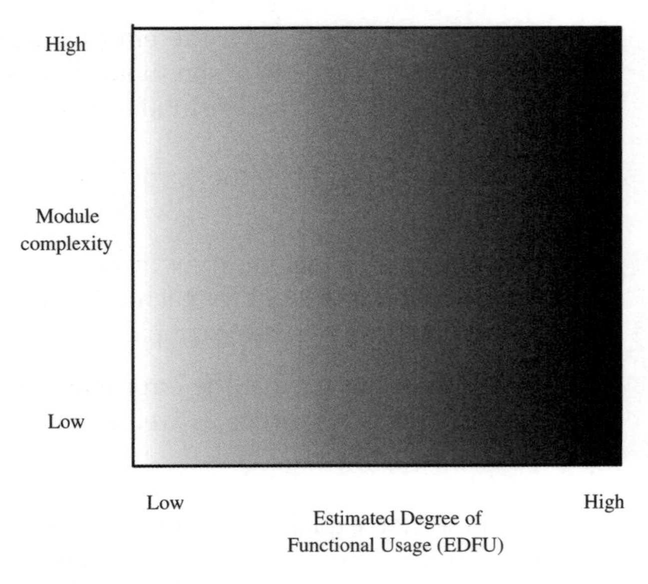

Figure 11-3. Defect criticality example 1.

However, even the features having a high EDFU (such as making a voice call with a mobile phone) are not automatically equally important from a testing point of view.

A software product normally contains some components with a complex architecture and other components with relatively simply architecture. The greater the complexity in a component, the more critical a defect is (see Figure 11-4), because such defects tend to have more interdependencies with other components. The complexity (see chapter 2) can be caused by two things, code complexity and the large number of shared resources (either because the component uses different resource files or because other components use its resources). More information on how to improve the focus of testing can be found in chapter 9.

11.2 Defect Analysis and Reporting

Testing ideally finds all defects that damage the product usability in the most used functions and guarantees that the most complex parts have been well covered. This sounds simple, but it still cannot be guaranteed. However, in most product programs the problem is not to find too few defects but to be able to fix them in a managed way. A managed way means fixing the right defects without causing regression.

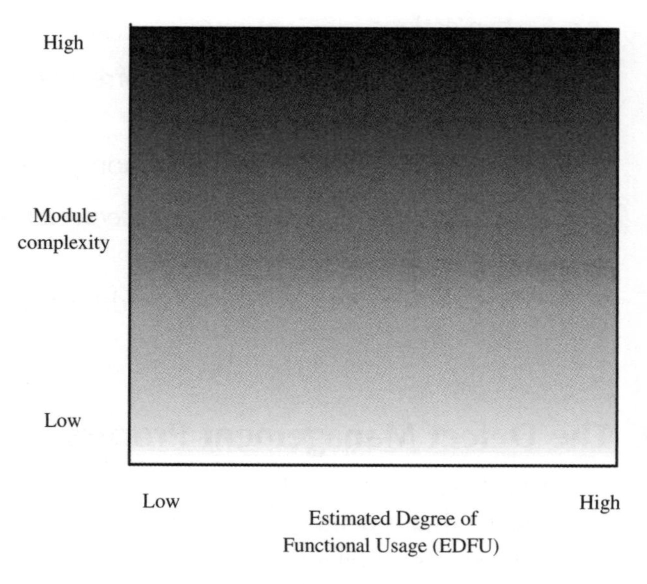

Figure 11-4. Defect criticality example 2.

Once the testing has checked the most important components and the most important defects have been discovered, it is time to start prioritizing and fixing the defects. In order to be able to do this, the program needs to have the following three things well understood and in place:

- a defect database
- a defect management process
- a defect priority description

11.2.1 Defect Database

Details of all defects should be stored in one place, where they can be managed and updated easily and whenever needed. Typically, this storage is either a shared file or database. The benefits of one well-planned defect storage area are:

- Impact analysis can be done faster when all entities are using the same database.
- Everybody can be informed of the existing failures.
- Fast reaction to errors entered into the database can be ensured.
- It will be possible to focus on the errors and to follow-up corrective actions.
- It can be ensured that defect correction is planned, implemented and verified.
- It speeds up the defect correction process.
- It gives information about defect status for product maturity estimations.
- Databases and processes should be consistent and not changed too often.

11.2.2 The Defect Management Process

Once discovered, a defect is reported into a common system. This means that a defect report exists. A defect report can have several statuses. A simplified set of statuses is shown in Figure 11-5.

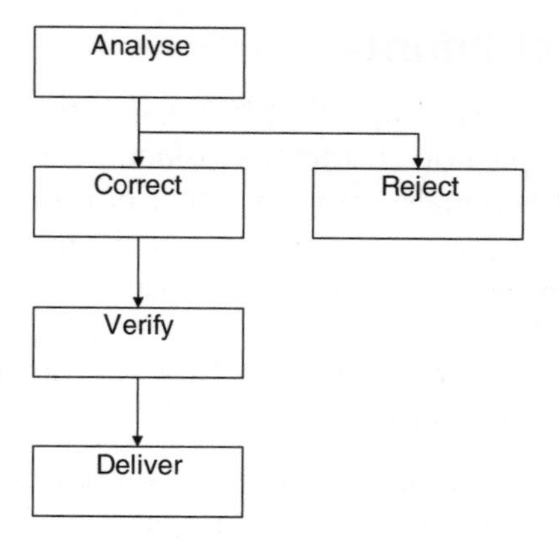

Figure 11-5. Simply defect report status lifecycle.

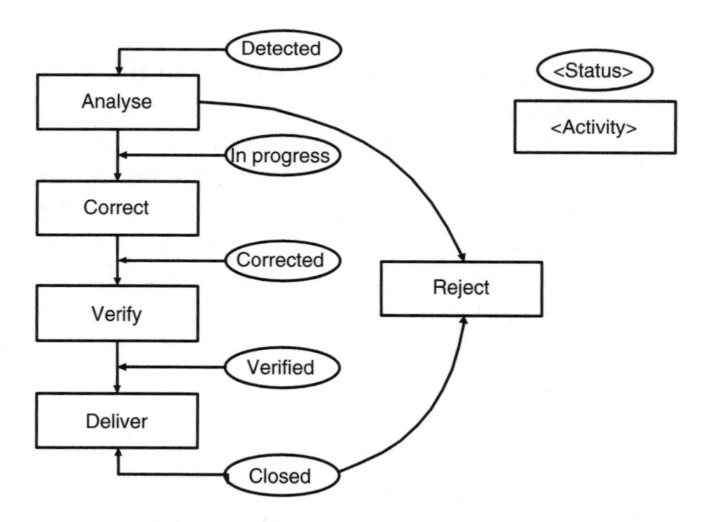

Figure 11-6. Defect process example.

The process can be considered as a road and the lifecycle tells where (in which phase) a car (the defect) is on that road. Once it is known what the defect status is, what happens next and who is responsible for the action are also known.

Figure 11-6 shows an example of a defect process with the activities and statuses. Status is a defect-specific indicator in the file or database that tells a person what is expected to happen to the defect next. Activity describes the phase of work around the defect. Activity needs to be assigned to a named person so that interested people can direct their questions to the right address.

11.2.3 Defect Priority

It is very important to be sure that those defects that have a major impact on customer satisfaction are corrected first. At the same time, it is equally important that non-business-critical defects in complex components are fixed last if time allows. Fixing priority order has three dimensions and all of them should be weighted equally:

- **Severity** describes how the defect impacts the system and how permanent the effect is.
- **Frequency** describes whether the failure occurs every time when using the piece of functionality.
- **Probability** describes how many customers have met with the problem and how likely the use case in question is.

Severity can have for example three levels: *high*, medium and low. *High* indicates that the product is unusable because of an error, or the software requires a heavy user interaction to recover from the failure (for example, removing the power supply). *Medium* indicates that the failure impact affects working with the product or recovery requires minor user interaction such as the system recovers when some button is pressed. *Low* indicates that the failure can be annoying but does not require any user interaction to recover from it. Such defects are for example UI issues, spelling errors and localization problems.

The scale of **frequency** levels used is very much product specific. In a mobile terminal, which can be assumed to be in use for a couple of hours per day, the frequency levels should be rather small (for example every minute, daily or weekly).

The **probability** should indicate the EDFU with the maximum frequency level.

Defect priority should tell how urgent it is for the defect to be fixed. There are different ways to describe the priority and one possible way is described here. A defect can be given one of the following four priorities: Show Stopper, Critical, Major and Minor. Each of these priorities is analysed below from three viewpoints: effect on customer, effect on business case and effect on R&D.

11.2.3.1 Show Stopper

Show Stopper is the highest priority. It is used if an error is really preventing/endangering something commercially important such as sales to a certain region or operator, ramp-up, the ability of the program to proceed or program milestones. Every defect having a show stopper status will have a severe effect on the customer, the business, R&D and the schedule and resources.

Effect on Customer

End users will not buy, or will return, the product if the error exists. End users will be very unhappy or angry about the usage of the product if the defect remains unfixed:

- **Severity**. The device is dead, needs flashing or rebooting to get it to function again. No workaround available.
- **Frequency**. Occurs very frequently, i.e. daily or even more often. This would have a severe impact on the FFR.
- **Probability**. The problem happens every time the basic functionality is used. A significant number of end-users experience the problem.

Effect on the Business Case

- There is no business for the product at all (major conflict with authority requirements) if these defects are not fixed before the product enters the market.
- There are hundreds of thousands of euros or more in lost revenue due to the error.
- The sales to certain regions or operators would be prevented.
- The progress of the program would be endangered.

Effect on R&D

- The project development depends on this functionality or fix.
- The project cannot go on until the error is corrected.
- Testing on a large scale would be prevented.
- Program milestones would be at stake.
- Open show stoppers in a late phase of the project will cause slippage in the project schedule.

Show stopper defects must be fixed as soon as possible and all resources should be used if needed in fixing show stopper defects.

11.2.3.2 Critical

Critical is the second highest priority. Priority Critical is used if an error is such that it does not prevent sales but has a major impact on customer satisfaction. Critical can also be used if an error causes problems to the program (e.g. prevents parts of the testing or endangers the milestones).

Effect on End-user

The end-user will buy the phone, but will be very dissatisfied with the product performance and functionality. The basic functionality is not OK. It is very likely that product will be returned to the service point. A minor problem that recurs all the time can be a critical error, as can an error with a heavy impact on the system, even if the error case is not likely to happen:

- **Severity**. Requires reboot or the device boots itself. Workaround may be available.
- **Frequency**. Occurs frequently. Frequency varies from hours to a week. The problem is likely to happen during normal usage, i.e. the MTBF is relatively high if this defect remains unfixed.
- **Probability**. Perhaps 5 per cent of end-users experience the problem.

Effect on Business Case

- The business and brand risk is unbearable.
- There are tens of thousands of euros in lost revenue due to the error.

Effect on R&D

- Product development and/or testing are partly prevented as a result of the problem.
- Open critical problems in a late phase of the project may cause slippage in the project schedule.
- Critical defects must be fixed as soon as possible, once no show stopper defects exist. However, solving the problem should not take up all available resources.

11.2.3.3 Major

Major is the third highest priority. Priority Major is used if an error causes serious problems but the use case is not common or the frequency of the error occurrence is rare.

Effect on End-user
End users will buy the phone, but they will be unhappy with the functionality of the product. If end-users find the error, only a few of the end-users will return the product.

- **Severity**. Very annoying behaviour of the device. Application shuts down itself. Workaround is available.
- **Frequency**. Occurs every now and then. In normal use, this failure is not very likely to happen, i.e. the MTBF is low even if this defect remains unfixed.
- **Probability**. Perhaps 5 per cent or fewer of the end-users experiences the problem.

Effect on Business Case
- Business and brand risks are significant.
- There are thousands of euros in lost revenue due to the error.

Effect on R&D
- There are effects only if there are a large number of major errors to be fixed before large-scale system testing starts.

All Major defects should be fixed before the system test phase, but there is currently no majr effect on development. Fixing should not require any new resource allocations.

11.2.3.4 Minor

Minor is the lowest priority level in this example. Priority Minor should be used in cases where the defect would hardly be noticed by the end-user or if it remains unfixed, it will not cause dissatisfaction with the product.

Effect on End-user
A Minor error is more a matter of taste or cosmetic in nature, so the end-user is not likely to notice it:

- **Severity.** Annoying behaviour of the device. Workaround is available.
- **Frequency.** Occurs very rarely. It may occur during heavy usage of the device, i.e. the MTBF would be very long if this defect remains unfixed.
- **Probability.** Only a nominal number of end-users experience the problem, not a normal use case.

Effect on Business Case
- There is neither brand risk nor significant risk of losing revenue as a result of the error.

Effect on R&D
- There is no effect on R&D.

Minor defects should be fixed if time allows and they should be resolved with the resources available at the time.

11.2.4 Defect Reporting

Even the very best defect reporting tool cannot replace the need for open communication between a tester and a developer. Neither can it replace a missing working procedure for defect handling (common rules) in the program. There will always remain things that the tester (the person who discovered the defect) does not put into the system either because he or she does not remember, because he or she does not consider them important or because it is difficult to explain them in a formal way. Such things can still be very vital for making a proper fix. Jason Yip opens this up in a very clear way in has article in the magazine *Better Software*.[1]

Writing only a formal defect report allows room for mistakes. These mistakes can cause extra delay in solving cases. A mistake in the use-case description can cause a developer to reject a critical defect report because he or she is not able to reproduce the defect. Therefore, a demonstration of the failure makes it easier to understand how the system behaves. As Yip describes in his paper:

> There is much tacit knowledge transferred in a conversation and demonstration (i.e. show and tell) that does not come across in a failing test case. For example watching the tester step through the problem allows the

observation of important details that the tester may not have thought useful to provide. Such details sometimes lead to the serendipitous discovery of related issues or even to the realization that the 'bug' is not actually a problem.[1]

Sometimes the discovered defect is actually a symptom of an underlying process problem, which will be left unknown unless a tester communicates this clearly to the developer:

Defects don't appear by themselves – they are injected. When a defect is detected, it may be an indication of an underlying process problem that will continue to inject additional defects. So the longer it takes to address the defect, the higher the likelihood that additional defects will be injected.[1]

A third fact to support open communication between tester and developer is that these two activities, being equally important, have very different natures. Development is constructive work, whereas testing is destructive work. Being human beings, it is understandable that, once the tester informs te developer that the code the latter has created is not working, the developer can easily become defensive. In such a case, it may be worth trying to introduce a human touch into the communication between these two roles, for example, making them sit close to each other and allowing them to get to know one another as a person. Regular meetings with both parties may also help to bring the needed consistency to the project.

11.3 Summary

Once testing has successfully discovered the most important defects in a product, programmers need to start fixing them one by one. This phase is often called the fixing period. In order not to create multiple new defects as a result of each fix, one needs to be sure what to fix, when to fix it and how to fix at. This chapter has introduced tools and processes to ease this important phase; professional defect handling is essential in the creation of a credible smartphone.

Chapter 12: Integration and Build Environment

S60-based phone integration needs to follow certain predefined steps. If the order is not followed, build creation fails. Such steps are introduced in this chapter. All middle-sized (and larger) software projects tend to set quite strict expectations on both Software Configuration Management (SCM) and build environment. This chapter introduces some general targets concerning configuration management.

12.1 Software Configuration Management

Whenever there is a need to maintain software, there is a need for software configuration management. Software maintenance can be both a long-term activity and a short-term activity. Long term means that a product needs to be reusable without there being major changes needed in the future. Short term means the activities within one product program, i.e. implementing the very first versions of modules and after that making the needed updates to the module in order to increase maturity. Every time a modification is made to a software file, then a check in and check out are needed. A proper software configuration management process contains as a minimum two things:

- rules for changing the code
- rules for the configuration management tool

12.2 Changing the Code

The larger the software under development, the more important it is to have good and reliable control over code changes. Following certain approved rules in the project enables this. Such rules are, for example:

- Keep all source code under version control.
- Use minor version numbers to distinguish between different versions of the same file and major version numbers to distinguish between different major versions of the project.
- Have named owners for each file.
- Control the access rights to the existing files.
- Have each potential change analysed before implementation.

In the state where all code exists (although the maturity of the code can still be low), in other words when the code-complete phase is reached, it becomes crucial to pay attention to making only managed changes that increase the stability. Sometimes a change correcting an important piece of behaviour creates several new defects. This is called regression. Most large software projects face regression at some point. However, if it happens too often, it can indicate loose change control in the program. On the other hand, if no regression is discovered at any point, the program should re-evaluate its testing efficiency.

At the point when a tester or programmer discovers a new defect, there is also a need for reverse engineering the case. Without having all changes recorded in one way or another, reverse engineering the situation back to the original version can be very difficult, if not impossible. Figures 12-1 and 12-2 show cases of reverse engineering.

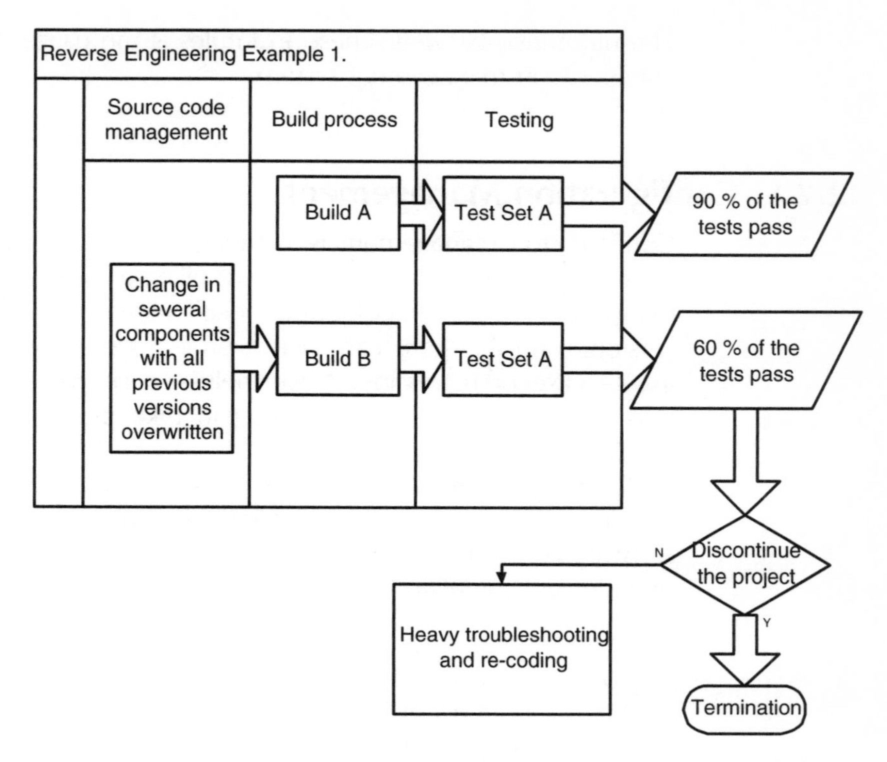

Figure 12-1. Reverse engineering example in an environment with no SCM.

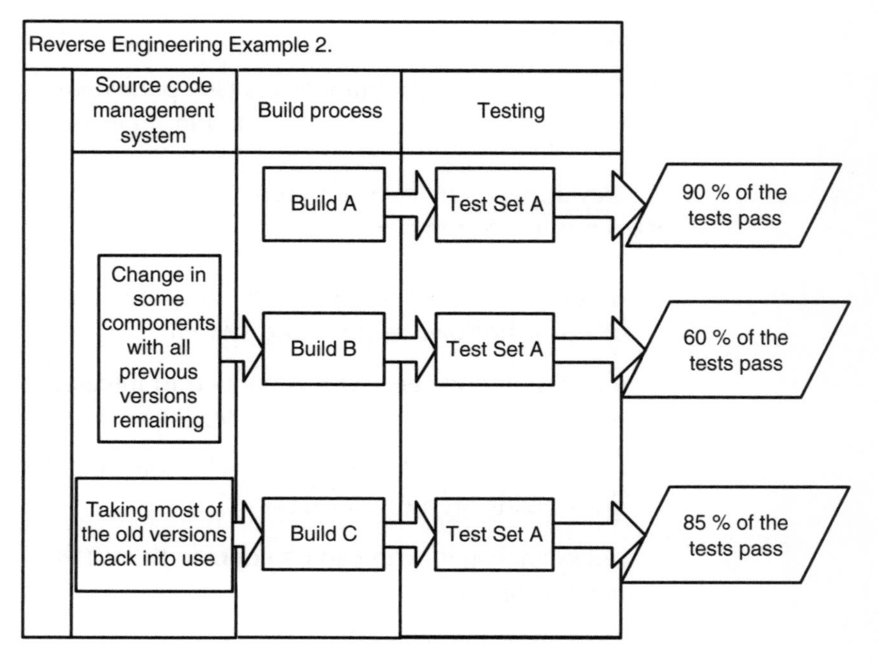

Figure 12-2. Reverse engineering example in an environment with SCM.

The difference between these examples is the usage of a software configuration management system.

12.2.1 Configuration Management

Configuration management is suitable not only for managing software configurations but also for controlling a wide variety of other things such as documentation, specifications and sub-systems. A version control system can save a project from complete disaster in the case of regression. Nevertheless, a simple system without rules on when and how changes can take place is not much of a help.

The program management should carefully plan moving the organization from a manually controlled procedure to the usage of a software configuration management tool. The following three phases provide one guideline for this transition:

1. Copy the currently followed manual process of teams integrating their code and transfer the completely integrated software into the selected SCM tool as one block.
2. Train each individual engineer to use the new SCM tool so that everybody checks in his or her own changes into the system.
3. The usage of team branches versus personal branches needs to be decided and personnel trained in their use.

Developers can have different roles in many SCM tools. Such roles are for example:

- developer, an ordinary programmer, who can store new versions of his or her own code in the system
- integration engineer, who is entitled to integrate the team's code into one sub-system for further testing and usage
- build manager, who integrates all code into the main line and does builds on a regular basis
- policy manager, who can can decide the variety of different submission policies

In many terminal programs the software is built of pieces from different sources. Sometimes these pieces are fully outsourced and the company is not using the same SCM tool as the program. There is a need for a procedure describing what, how and when such pieces of software are to be integrated into the base.

12.3 Build Environment

The S60-based smartphone software needs to be built many times within each product program. As the software is rather significant, it sets some limitations and guidelines on both delivery structure and program processes. This section introduces all the hardware requirements as well as describing one process the customer can follow.

12.3.1 Delivery Structure

The program needs to acknowledge that each S60 delivery includes many things and has a predefined structure. The content of an example delivery is:

- S60 source codes
- sources of the adaptation layer stubs
- integrated Symbian OS sources for the platform
- S60 binaries
- release note
- change log
- structure changes log
- build tools

The S60 development environment consists of development computer software and the S60 release on an otherwise empty drive. The S60 drive contains the platform plus the Symbian OS combination and an engine (adaptation, base port and modem software), which should come from the phone program.

The directory structure of the development environment is the following:

```
\S60 - Pure S60 sources
\src\beech\generic
\src\common\generic - Symbian OS release for the S60
\adaptation - Adaptation layer stubs
\epoc32
    \epoc32\release - Release binaries for different platforms (ARM4,
        Thumb, Wins and WinCW)
    \epoc32\Rom - Rom creation kit
    \epoc32\data - emulator data (background image, .ini file etc)
    \epoc32\gcc - Compiler for target builds (thumb, arm4)
    \epoc32\include - All header files
```

The S60-related components should have the following structure:

```
\S60\About
    \S60\Data\ - resource files
    \S60\Group\ - build info files
    \S60\Inc\ - Header files
    \S60\Src\ - cpp files
```

S60 build tools play an essential role in customer programs' build procedures. The build managers need to be well trained in using these tools. The build tools are the Rom image creation tool, the Symbian decentralization tool and the Symbian basic build tools:

The Rom image creation tool S60Rom.cmd generates an image for the hardware. There are two section Rom images in S60:

- Core Image contains executable code (EPOC SW) and all the epoc resources excluding localized bitmaps.
- Variant Image contains language resources and all such software that has country- or operator-specific settings.

12.3.2 Build Process

The build process should be able to provide a guide to resources throughout the product program when it comes to having the latest version of the software available whenever needed. The frequency of building should be accommodated to a project's requirements. As mentioned in chapter 2, a project's need for a new build varies over time along with the overall stability increase. For a build cycle to be flexibly changeable, a process needs to be defined and strictly followed. Figure 12-3 shows a possible build process in a S60-based phone program.

Localization needs to be done whenever a product is targeted at non-English markets. There are two different types of localization:

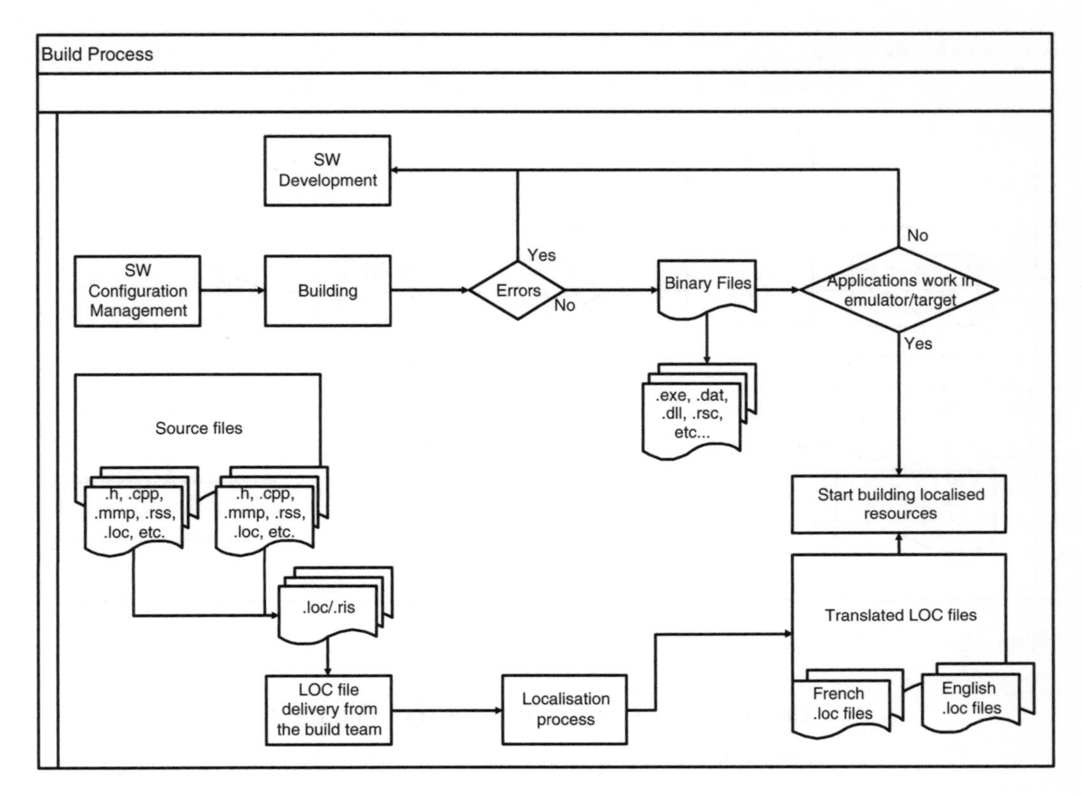

Figure 12-3. Build process.

- language variants, such as the Chinese variant
- cultural variants, in which some features may need to be removed before shipping the phone in that country

Figure 12-4 shows a localization process in a product program.

12.3.3 Build Tools

Building a S60-based software package requires the use of a certain set of tools. Some of these tools are Symbian based and some platform based.

There are two types of Symbian basic build tools, the decentralized and basic build tools.

Decentralized means that the tool can be opened by a command. For example, Genxml.pl generates xml input for the build server, i.e. generates Buildserver.pl and buildclient.pl files.

Basic build tools are:

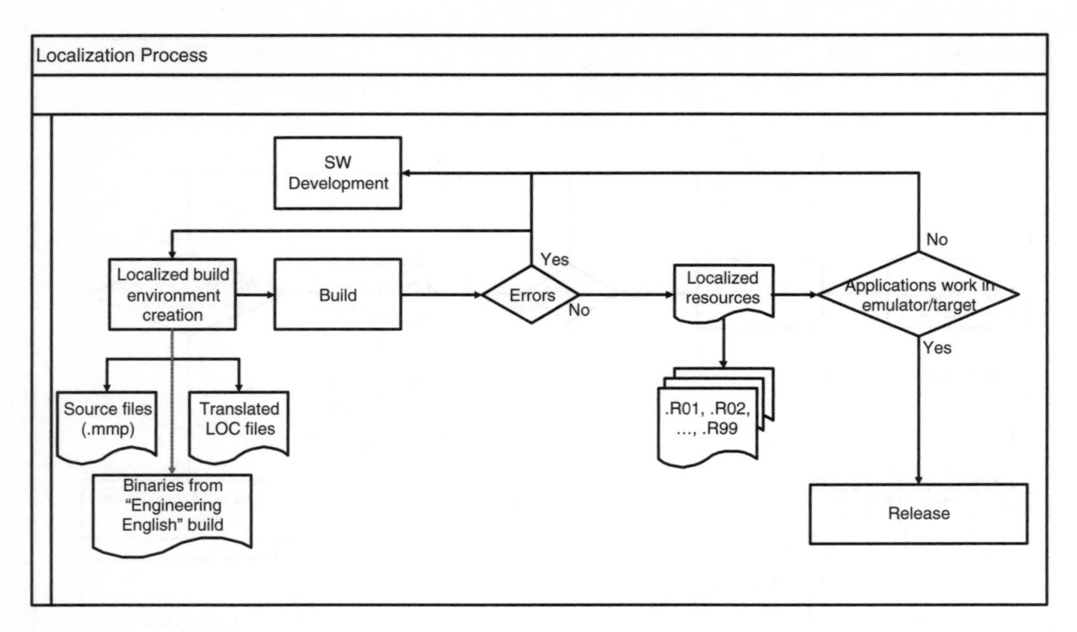

Figure 12-4. Localization process.

- Bldmake bldfiles, which creates the abld.bat-file
- Abld build, which runs the abld makefile, abld export, abld library, abld target, abld final (i.e. builds the component in directory)
- Abld thumb, which creates only thumb binaries
- Abld build –w, which lists all that has been built in the directory
- Abld build –c, which checks that every component in the directory has been built

There are two kinds of S60-based build tools, the basic build tool and a set of Rom image creation tools

1. Build_S60.cmd, which cleans and builds S60 binaries from the input files specified in build scripts Averell30_bld.txt and Averell30_rc.txt.
2. Rom image creation tools.
3. S60Rom.cmd, which generates image for the hardware. In S60 the Rom images are two section rom-images:
   a. Core Image contains executable code (Epoc sw) and all the Epoc resources that are not localized plus the bitmaps.

   b. Variant Image contains language resources and all software country or operator variant settings. Variant image is normally flashed in Label-place in the production line.

4. The localized_sc.iby file contains the information on all localized resources needed to be included in rom-image (the rule is that if the text is seen in the UI it needs to be localized).
5. The program itself defines how many languages and variants it needs to implement.

In addition, the use of some other tools for checking the building success is highly recommended. Such tools, which can carry out some level of build sanity check, are especially valuable in the busiest phase of the program. An example of such a tool could be one that analyses the S60 build against the master compilation oby file. It reports possible missing files according to the specifications in configuration file. This kind of tool is normally a shell tool.

Another valuable tool is the AppDep-tool, which resolves which libraries a certain component uses as well as which components are utilizing this component.

12.4 S60 Integration

12.4.1 Stage 1

Stage 1 is a kind of backbone to the rest of the integration activities. If it is not successfully completed, there is no reason to continue the procedure. Stage 1 contains the following steps:

12.4.1.1 Step 1: Successful Boot to Textshell

The first step can be considered as a base test for Baseport delivery and it verifies text shell image creation and successful boot with S60 clean build and BSP-delivery. The only things in the ROM in this step are the started operating system, the loaded required device drivers and the launched textshell via the operating system's WSERV.

It is highly recommended that this step is executed every time before any S60 integration into the hardware. This is because it already says a lot about the first phases of boot in terms of the functionality, starting the lower-level implementations, which are the base assumption for S60. It should anyway be kept in mind that there are different implementations of some drivers etc. for the textshell and for the graphical UI. This phase is not part of the S60 integration process but more likely a pre-requisite for it.

The current S60 implementation includes stub components for adaptation layer drivers etc. These stubs enable the software to boot to the application shell with the minimal driver set.

12.4.1.2 Step 2: Simple Application and Launch via WSER

The recommended way to start integration of S60 into hardware is to launch some simple application (for example, the calculator) via WSERV after a successful boot to textshell. When this is done, all the required components need to be in ROM and some dependencies need to be removed since only small part of S60 is taken into use. In addition, the integration team may find it useful to enable DLLResource loading debug-prints.

12.4.1.3 Step 3: Starter Integration and Calculator Launch

Now, if previous steps were successfully completed and the hardware contains a keyboard, more and more software can be integrated into a build. The following sub-steps are recommended at this stage:

1. Remove changes made to WSERV in previous step.
2. Modify Starter's start-up list so that it will take care of starting the calcsoft and all needed components.
3. Take all required files into ROM.

12.4.1.4 Step 4: Complete the S60 Boot

This step can be achieved by debugging the boot process to see how the previous steps succeeded.
The success of Stage 1 is often called milestone 1.1, which is discussed in chapter 3.

12.4.2 Stage 2

Stage 2 is aimed at integrating the adaptation components and, if it is successfully completed, a simple voice call can be established. After successfully completion of stage 2, milestone 1.2 can be considered as having been reached.

12.4.3 Stage 3

Data connections are the next things to include in the build. The order of different connections is recommended to be as follows:

1. CSD/HSCSD data connection integration.
2. GPRS connection.
3. WCDMA connection.

After stage 3 has been successfully completed, milestone 1.3 can be considered as having been reached.

12.4.4 Stage 4

Local connectivity is next targeted for inclusion into the build. After stage 4 all connectivity protocols that the phone supports, such as Bluetooth, Infrared and USB, should work.

12.4.5 Stage 5

Finally, and yet importantly, the components implementing multimedia such as the camera and audio should be included into the build.

12.5 Summary

Creating a software build out from the source code can be demanding. What complicates this in the world of S60 is the platform architecture combined with the size of the system. This chapter has explained what preparations are needed for the build and how a build is created. In addition, the integration order of the components has been explained.

Appendix A: Examples of S60 Devices

S60 is without any doubt the world's leading smartphone platform. It has been delivered to consumers within a variety of different mobile devices. Overleaf is a collection of Nokia S60 devices in the market at the time of writing.

S60 Smartphone Quality Assurance Saila Laitinen

Figure A-1. The Nokia N80 smart multimedia device is a 3G world phone with EGSM 850/900/1800/1900 and WCDMA 2100 for Europe, Africa and APAC regions and EGSM 850/900/1800/1900 for China. A three-megapixel digital camera, email, digital music player, personal organizer, game console, UPnP and WLAN connectivity, makes the N80 Nokia's most advanced all-in-one device yet.

Figure A-2. The Nokia N73 is a stunning multimedia computer with powerful photography features and integrated stereo speakers with 3D sound. In addition to providing the standard range of Nokia N series multimedia experiences, the Nokia N73 includes a 3.2 megapixel camera with Carl Zeiss optics, auto focus, two-way video call capability and MPEG-4 Video capture at 15 fps.

Figure A-3. The Nokia N93 features a 3.2 megapixel camera, Carl Zeiss optics, 3× optical zoom and digital video stabilization. Create DVD-like videos at 30 frames per second with MPEG4 technology and share them on the 2.4″ display. For a big screen experience, connect the N93 to a compatible TV using direct TV out connectivity or via Wireless LAN and UPnP technology. The N93 also features a digital stereo microphone, music player and FM stereo radio, dual mode WCDMA/GSM and triband GSM coverage on up to five continents (EDGE/GSM 900/1800/1900 + WCDMA 2100 MHz networks).

Appendix B: Glossary

API: Application Programming Interface, a set of services an application developer can utilise when implementing applications on top of a platform.

Back bone testing: the core components are implemented and tested first. Only once a stable enough set of core components is obtained, are the components that utilize these core component services integrated and tested.

Backward compatibility: if an application implemented with the help of SDK release *x* runs on a device based on platform version $x + 1$, that device is backward compatible with the application.

Baseline: the release that has been integrated as a whole in the customer device software.

Base porting: the exercise of adapting the Symbian kernel to particular hardware.

Basic Acceptance Testing (BAT): a small sub-set of all functional test cases of the platform BAT cases is run on every single release; the result indicates whether a particular release is mature enough for further testing.

Big bang testing: the entire code is tested at one time.

Binary compatibility: all versions of one platform conform in terms of the API set.

Black-box testing: testing activity accomplished without knowing the code's internal architecture. Black-box testing approach can be applied in all testing phases (module, integration, functional, non-functional and interoperability testing).

Bluetooth (BT): an industrial specification for wireless personal area connectivity. In smartphones the user can connect to external devices such as PCs or printers by establishing a Bluetooth connection.

Bottom-up testing: a testing approach where the system architecture is built from the bottom. In other words, the components that create the base of the system are integrated and tested first.

Camcorder application: a digital audio video encoder and player application in a device.

Cellular Telecom Industry Association (CTIA): an international organization representing all wireless sectors and providing non-profit memberships to service providers, manufacturers, wireless data and Internet companies, as well as other contributors to the wireless universe.

Client provisioning: managing applications and content for networked devices. All mobile devices are unique in one way or another and have different features enabled. This means that application and content providers have to know the device specific capabilities before loading an application into a device.

Code complete: all needed software has been implemented and is ready to be integrated.

Code Division Multiple Access (CDMA): a technique in which radio transmissions using the same frequency band are coded in such a way that a signal from a certain transmitter can be received only by certain receivers.

Code reviews: a testing technique in which highly competent engineers print out the code and read it together to find illogicalities and defects.

Constructive testing: testing in which the tester tries to show that the system works. The tester is not interested in discovering the defects. Constructive testing is often based on real use cases.

Conversion Description Language (CDL): an interface allowing access to the layout data. This layout data is based on the S60 look-and-feel specifications. Layout data for an application can be stored by utilizing the CDL interface.

Customer program: any device program that is based on the S60 platform. It can be from Nokia or from some other device manufacturer licensing the platform from the Mobile Software (MSW).

Data-driven testing: a testing technique aiming to find defects in which certain data is wrongly processed. It focuses on every data area of interest in a product.

Dataflow driven testing: a testing technique focusing on problems in component interfaces.

Defect estimation: a method to evaluate the success of testing in a program.

Defect frequency: a metric indicating how often a defect occurs in a product. For example, with a defect frequency of 1/2, the defect occurs every second time the user uses the device in a certain way.

Defect probability: a metric indciating how probable it is that the user comes up against a defect when using the product. It describes how many customers meet the problem and how likely the use case in question is.

Defect seeding: a method for verifying testing success and efficiency. If, for example, 100 defects are seeded into a system and during testing 70 out of these 100 defects plus 70 extra defects are discovered, one can assume that the system contains a further 30 unknown defects.

Defect severity: describes how a defect impacts the system and how permanent the effect is.

Denial of Service attack (DoS): a security hack often caused by an extensive load on a product. This load can cause the product to block all its services and functionalities.

Destructive testing: a testing approach in which the intention is to break the system under test. In other words, testing showing that the system does not work in the way that it is supposed to work.

Digital Rights Management (DRM): a rights management system that ensures that content can only be used when the relevant conditions, determined by the copyright owner, have been met.

Dynamic Link Libraries (DLL): a function library that can be loaded into memory once and called by one or more applications so that the operating system dynamically resolves at run time the entry points, or the addresses, of the routines that are called.

Enhanced Data Rates for Global Evolution (EDGE): a radio interface modulation technique that increases HSCSD (high-speed circuit-switched data) and GPRS (general packet radio service) data rates.

Estimated Degree of Functional Usage (EDFU): a numeric value indicating how probable it is that an end user uses a particular service/functionality/application in a product. For example, a value of one indicates that the product is never used without the usage of this service/functionality/application.

European Telecommunications Standard Institute (ETSI): European organization that produces standards that are applied and accepted in the area of telecommunications.

Event driven testing: an approach that tries to reveal incorrect handling of events.

Extreme programming: a development process depending on clear communication, simplicity, feedback and courage. These four topics are ensured by having programmers sitting in pairs while coding.

Feature phone: a common term for a phone that has a relatively simple but effective, proprietary software environment based on a real-time operating system (RTOS).

Field Failure Rate (FFR): a measure of product quality and reliability. It indicates how soon after the product being available the very first defect is discovered by the end user.

Forward compatibility: if an application implemented with the help of SDK release x runs on a device based on previous version $x - 1$ of a platform, that device is forward compatible with the application.

General Packet Radio Service (GPRS): a GSM data transmission technique that transmits and receives data in packets. GPRS offers a permanent connection between the wireless device and the network.

Global Certification Forum (GCF): an organization that aims to maintain confidence in new mobile wireless terminals by means of product certification. Manufacturers are encouraged to certify their products in accordance with a series of agreed criteria.

GSM circuit switched data: data that is transferred via a circuit-switched network using the Global System for Mobile Communications (GSM).

GSM high-speed circuit switched data: a data transmission connection that is few times faster than the GSM data connection. It uses multiple channels for data transmission.

High-water mark definition: an operation defining the maximum load a product can handle with a predefined service level.

Independent software vendors: suppliers that are independent developers and resellers of products based on a particular computer hardware or operating-system platform.

Java 2 Micro Edition (J2ME): a Java application environment that forms a framework for the deployment and use of Java technology in the post-PC world.

JUnit: a freeware testing framework used by developers who implement unit tests in Java.

Lead environment: the device program used for development and testing purposes in a platform development program.

Licensee: the person or company licensing the rights to use the licensor's proprietary application.

Logic driven testing: a testing technique aimed at identifying all incorrect handling of the logic in a product.

Look-And-Feel (LAF): (1) the effect that the appearance and functions of a program's user interface have on the user; (2) user interface guidelines for platform application developers.

McCabe's cyclomatic complexity: a measure of the number of linearly-independent paths through a program module. It is the most widely used member of a class of static software metrics. Cyclomatic complexity may be considered a broad measure of soundness of and confidence in a program.

Mean Time Between Failures (MTBF): an expected time between failures. For example, the expected operating time between two consecutive system failures of a unit.

Mobile Software (MSW): an organization within Nokia providing the S60 platform to all customer programs.

Multimedia Card (MMC): a flash memory card standard. Typically, an MMC card is used as a storage medium for a portable device.

Open Mobile Alliance (OMA): an industry forum for developing market-driven, interoperable mobile service enablers.

Platform security: a feature of S60 3rd edition ensuring, for example, data caging.

Product Creation Community (PCC): community of technology integrators and other companies potentially interested in, and capable of helping out, the customer product program in making a phone.

R&D quality: a product quality level during development.

Real-Time Operating System (RTOS): an operating system that performs data processing in real time, or at least with a low delay time.

Reference hardware: a semi-smartphone with basic S60 functionality. Reference hardware is used in S60 testing and it can be used as a base in a S60-based device program.

Software Development Kit (SDK): a set of programming tools for creating applications and enhancing the use of certain software. In S60 it is a product that provides tools, documentation and end-to-end application examples that support the development of applications, mediations and adaptations on top of the platform.

S60 ecosystem: the entire set of different players in the S60 platform community. It contains PCC members, technology integrators, third-party developers, product programs and the platform organization.

S60 third edition: the S60 release 3.0.

Smartphone: an electronic device that integrates the functionality of a mobile phone and a personal digital assistant (PDA) or other information appliance. A key feature of a smartphone is that additional native applications can be installed on the device. One significant characteristic is its multi-processing capability.

Source compatibility: an application or client program can be rebuilt without the need to modify the program.

State-driven testing: a testing approach aimed at finding all incorrect state transitions.

Static analysis: a testing technique performed without actually executing programs built from that software.

System Under Test (SUT): the set of code being tested.

Technology Compatibility Kit (TCK): a suite of tests, tools and documentation that determines whether or not a product complies with a particular Java™ technology specification.

Testability: the degree to which a system or component facilitates the establishment of test criteria and the performance of tests to determine whether those criteria have been met.

Testware engineering: a full-life-cycle process that must be initiated when the project begins to be maximally effective.

Third-party developer: an independent entity innovating on top of some platform.

Three-D-rule: a process for testers. It is based on the three words; Decide, Define and Distribute.

Top down testing: a testing approach in which the system under test is built from the top to bottom. For example, the user interface is created and tested first and only after that is successful are the rest of the underlying components integrated.

TRUE testing: a beta testing technique for verifying that the product functions in real use by using end-users to detect errors. It requires the involvement of volunteers who agree to use the product on a daily basis. TRUE testers need to provide feedback to the product program.

Universal Serial Bus (USB): a plug-and-play interface between a computer and a compatible add-on device, such as an audio player, joystick, keyboard, phone, scanner, digital camera or printer. With a USB, a new device can be added to a compatible computer without having to add an adapter card or even having to turn the computer off.

White-box testing: a testing technique in which the tester has a clear understanding of how the system under test has been structured.

Wireless Application Protocol (WAP): an open, global standard for total mobile solutions, including communication between a mobile handset and the Internet or other computer application.

Appendix C: References

Chapter 4: Binary Compatibility

1 Szilagyi, Sandor (2003) *Binary Compatibility Theory Training Material*. Nokia/MSW [available to S60 Licensee]

Chapter 5: Certificates and Standards

1 More on Java available at <http://java.sun.com/j2me/>

2 More on BT available at <http://www.bluetooth.com>

3 More on MiniBae available at <http://www.beatnik.com>

4 More on T9 available at <http://www.t9.com/>

5 More on VeriSign available at <http://www.verisign.com/>

6 More on Baltimore available at <http://www.baltimore.com/unicert/technology/wtls.asp>

7 More on Entrust available at <http://www.entrust.com/>

8 More on USB available at <http://www.usb.org>

9 More on InfraRed available at <http://www.irda.org>

10 More on MMC available at <http://www.mmca.org/>

11 More on SyncML available at <www.syncml.org>

12 More information on ETSI available at <www.etsi.org>

13 More information on CDMA available at <http://www.cdg.org>

14 More on RTTE1999/5/EC available at <http://europa.eu.int/comm/enterprise/rtte/dir99-5.htm>

15 More on RoHS available at <http://europa.eu.int/comm/environment/docum/00347_en.htm>

16 More on FCC available at <http://europa.eu.int/comm/enterprise/automotive/directives/vehicles/dir70_156_cee.html>

17 More on PCS available at <www.ptcrb.org>

18 More on China Approvals available at <http://www.mii.gov.cn>

19 More on GCF available at <http://gcf.gsm.org>

20 More on CTIA available at <http://www.ctia.org>

21 More on ELSPA available at <http://www.elspa.com>

22 More on ISO available at <http://www.iso.org>

Chapter 6: What Quality Means

1 Dean, James W. Jr and Evans, James R. (2000) *Total Quality*. South-Western College Publishing.

2 Kingsley, Kimberley (2005) *A Foundation of Trust*. American Society of Quality/Quality Progress.

3 *Six Sigma Excellence Brochure* (2004) American Society for Quality.

4 Kim, Seong-Ho, Yoon, Yeo-Han and Zeon, Gyu-Tae (2004) 'Combine Quality and Speed' *Six Sigma Forum Magazine* available at <www.asq.org>.

5 Larman, Graig and Basili, Victor R. (2003) 'Iterative and Incremental Development: a Brief History' *IEEE Computer Science* **3**, 0018-9162.

Chapter 7: Stumbling Blocks

1 See <http://www.nasa.gov/centers/ames/research/technology-onepagers/risk-analysis.html>.

2 Construx Software Builders, Inc. (2000–2001) '10 Keys to Successful Software Projects' available at <http://www.issre2001.org/10KeysToSuccess.pdf>

3 Tiwana, Amrit and Keil, Mark (2004–2005) *Programming Languages* **2**(9).

4 Boehm, Barry (2002) *Software Risk Management* COCOMO/SCM Forum # 17.

5 Softstar Systems (1986–2005) Available at <http://www.softstarsystems.com/overview.htm>.

6 Pressman, Roger S. (1997). *Software Engineering – A Practitioner's Approach*, Fourth edn. McGraw-Hill.

Chapter 8: Platform Testing versus Platform-based Phone Testing

1 ISEB Practitioner Certificate in Software Testing. ISEB 7925-2. British Computer Society.

Chapter 9: Testing as a Tool

1 Craig, Rick (1995) *Software Quality Engineering*. Nokia Research Center.

2 'Agile Software Development'. Wikipedia the free encyclopedia, Wikimedia Foundation, Inc. Available at <http://en.wikipedia.org/wiki/Agile_software_development>.

3 Schaefer, Hans (1999) *SOFT-ED: Become an Expert in software testing*

4 See <http://www.testingstandards.co.uk/>.

5 See <http://en.wikipedia.org/wiki/Unit_testing>.

Chapter 11: Defect Analysis

1 Yip, Jason (2005) 'I don't Want a Bug Report-I'd Rather We Talk' *Better Software*, July/August.

Appendix D: Further Reading

Amler, Scott (2002) *Agile Modelling*. New York: Wiley.

Beizer, Boris (1983) *Software Testing Techniques*. New York: Van Nostrand Reinhold.

Beizer, Boris (1984) *Software System Testing and Quality Assurance*. New York: Van Nostrand Reinhold.

Beizer, Boris (1995) *Black-Box Testing: Techniques for Functional Testing of Software and Systems*. New York: Wiley.

Crosby, Philip B. (1980) *Quality is Free*, Mentor.

Evans, James R. and Dean James W. Jr. (2000) *Total Quality, Management, Organization and Strategy*. South-Western College Publishing.

Gilb, Tom (1988) *Principals of Software Engineering Management*. Addison-Wesley.

Gilb, Tom and Graham, Dorothy (1993) *Software Inspection*. Addison-Wesley.

S60 Smartphone Quality Assurance Saila Laitinen

Harrison, Richard (2003) *Symbian OS C++ for Mobile Phones, Volume One*. Wiley.

Myers, Glenford (1978) *The Art of Software Testing*. New York. Wiley.

Schaefer, Hans (2001) *Software Testing Days - Training Material*, see www.soft-ed.net.

Index

Page numbers in **bold** refer to the glossary.

J

K

L

M

N

O

P

V

W

X